Management Division

Presentations at the 2009 AIChE Annual Meeting

Nashville, Tennessee, USA
8-13 November 2009

ISBN: 978-1-61567-929-4

Printed from e-media with permission by:

Curran Associates, Inc.
57 Morehouse Lane
Red Hook, NY 12571
www.proceedings.com

Some format issues inherent in the e-media version may also appear in this print version.

Printed by Curran Associates, Inc. (2010)

For permission requests, please contact AIChE
at the address below.

AIChE
3 Park Avenue
New York, NY 10016-5991

Phone: (203) 702-7660
Fax: (203) 775-5177

www.aiche.org

TABLE OF CONTENTS

Managing Innovation in ConocoPhillips R&D

James Scinta, ConocoPhillips, Bartlesville, OK
Rena J.Armstrong, ConocoPhillips, Bartlesville, OK

ConocoPhillips has strived to maintain a robust Research and Development portfolio since the merger of Conoco Inc and Phillips Petroleum Company in 2002. Various approaches have been launched to encourage idea generation, project execution, and the conversion of ideas to commercial success. In recent years these approaches have settled into two different systems for funding research based on the development time frame.

Long term research ideas are relatively unrestricted however they do need to fit into the overall strategy of the corporation. Initial funding for this work usually comes from special funds at the discretion of senior management. Often times special research funding is set aside for research and development in areas of future interest to the corporation. There is no return expected for this type of research investment in the five year budget window.

A more rigorous system for management of nearer term research projects has been developed. The Corporate Technology Development (CTD) program was initiated to increase the development and implementation of refining improvement projects. These projects are expected to return value to ConocoPhillips' refining network within the next one to five years.

The CTD program was initiated with a call for ideas in selected areas that were identified by management. Ideas were evaluated by a joint panel with representatives from R&D and experienced refining technical personnel. Ideas that were deemed sufficiently promising were awarded limited funds to further develop the idea. These ideas were then further evaluated and program funds were awarded. Every awarded idea showed a reasonable probability of technical success relative to the funds being risked and included a plan of where and how this idea could be implemented in at least one refinery. Over the last five years this program has returned sufficient value to refining that management has increased the funding for this type of research.

The backbone of both systems is a two page form. The form contains a concise description of the project, list of key deliverables with deadlines, project budget, names the principal investigator and lists key contacts and people who must be informed about the progress of the project. It also contains the value proposition for the project and a plan for implementation of the project in the refining system. Success and failure criteria are also clearly stated. The information on these forms is tied into ConocoPhillips' performance management system so that researchers' performance is matched with project expectations.

Conquering Innovation Fatigue in Corporations: Helping Chemical Engineers Reach Their Innovation Potential

Jeff Lindsay, Innovation Edge, Neenah, WI

Cheryl Perkins, Innovation Edge, Neenah, WI

Mukund Karanjikar, Technology Holding LLC, Salt Lake City, UT

Chemical engineers have been among the most productive innovators of the past several decades, with broad and flexible skills that can deliver creative solutions to numerous problems. However, innovation productivity of engineers in corporations is often suppressed by a series of sometimes invisible or silent innovation killers. Innovators and management need to understand these "innovation fatigue factors" in order to find solutions or workarounds and help creative engineers realize their potential. We will discuss several concepts, including the "Horn of Innovation" paradigm, guidance on creating a healthy innovation culture, tips for strengthening the "will to share" in your innovation community, and recommendations for engineers facing internal innovation barriers.

See more of this Session: Managing Innovation
See more of this Group/Topical: Management Division

Satellite R&D: A New Model of Innovation

Mukund Karanjikar, Technology Holding LLC, Salt Lake City, UT

Topline growth is becoming a significant challenge in today's competitive world. New products and technology can give you the required edge. However, in energy and process industry, the product or technology development cycles tend to be significantly long and very innovations see the light of the day. One major reason for this is 'valley of death' for R&D projects.

This paper presents an overview of 'innovation catalysts',a breed of firms that thrive on making R&D as their core function and providing the bridge required for R&D projects to pass beyond a critical stage. A satellite model of R&D will be discussed. Using such a model companies can overcome the valley of death and launch more new products and technology.

See more of this Session: Managing Innovation
See more of this Group/Topical: Management Division

The TIC-TAC-TOE STRATEGIC PLANNING

Jack Hipple
Principal
Innovation-TRIZ
Tampa, FL

Abstract

When we study the history and trends of technology development across a wide spectrum of technologies, several things stand out. First, there are similarities in how difficult problems are solved in different areas of science and technology. We see the same principle of increased dynamism used in many mechanical systems to improve comfort as well as in the pricing of airline seats based on the then current fare yield and seats currently unfilled. Second, the evolution of technology also has followed certain predictable, though sometimes discontinuous steps. One of these is the field transition line we see in the evolution of communication "systems". A long time ago, we beat on drums (a mechanical/acoustical field). Then we discovered smoke (thermal field) and subsequently writing (chemical field). Telephones (electrical field) followed by cell phones (electromagnetic field). This same field progression can be seen in chemical reaction technology, cooking, indicating devices, and numerous other systems. Thirdly, product and business development evolves in such a way as to integrate functions and systems into their super-systems. For example, we can look at all the functions now incorporated into what used to be fax machines and TV remote controls, not to mention hotel keys and cooking systems.

These trends have been defined within the TRIZ (Theory of Inventive Problem Solving science) and can be pro-actively used in strategic product, engineering, and business planning in a very simple way through the use of a simple tic-tac-toe box. As we move from left to right in this box, we follow the lines of evolution mentioned above. Bottom to top visually describes the upward system integration. Systems inevitably move into the upper right hand corner of such a diagram.

1. Patterns and Trends of Evolution

"TRIZ" (Theory of Inventive Problem Solving) is a problem solving and strategic tool set and process based on the study of the top few percent of patents that illustrate breakthrough principles. TRIZ shows that there are a limited number of principles and patterns that apply across a wide range of products and technologies.

Several of the many patterns and tools contained within TRIZ are:

A. Products and systems are integrated into their super-systems over time.
B. Products and systems become more dynamic and responsive over time
C. Products and systems progress along a line of field evolution: mechanical, thermal, chemical, electronic, electromagnetic

This paper will review these lines and show a simple process for visualizing and using these concepts pro-actively in business and product development.

2. Super-system Integration

Think about the office machines and communication devices we use. We used to have telephones, copy machines, fax machines, and computers as separate and distinct products. Over the past 20 years, we have seen copiers and fax machines integrated into one unit that is also capable of copying. Our I-phones can now not only call, but surf the web, find directions (replacing map printing), and obtain instant sports updates (replacing the newspaper). These “replaced” products and services are now becoming extinct. This predictive line forecasts that any product or service will ultimately be absorbed into its super-system. Even the evolution of “super” stores such as Super Target and Wal-Mart Super Centers illustrate this in a commercial sense. Not only groceries, but personal services, restaurants, drugs, and pharmacies are in the same building as the grocery store. In chemical engineering unit operations, we see reactive distillation columns beginning to replace separate reaction and separation systems. We see sophisticated process control allowing multi-step synthesis in a single vessel as opposed to multiple reactors. We also see multifunctional product development as illustrated by Behr’s most recent paint product where the primer and final coat are integrated into one product, eliminating the first step in the painting process. On the business side we see upward integration of businesses in large corporations controlling and investing in both upstream and downstream businesses. We see multi-functional cleaning products where there used to be separate ones for each stain or clean up chore.

3. Dynamism and Responsiveness

Think about the simple examples we see in everyday living. Airline fares change daily based upon load factors, competitive activities, seat location, and date of booking. In California, toll road fees change as a function of time of day and current auto traffic. This would have been unheard of in the early days of the Pennsylvania turnpike. We see the same behavior in cruise ship cabin pricing as well as chemical pricing. In the case of pricing, we are talking about responsiveness. In the case of airline seat selection, we are collecting more money from travelers who prefer exit row seats, making the fare structure dynamic as opposed to static. In the automotive industry we see continuously variable transmissions and automatic radio volume control based on car speed as other examples.

In the chemical world, examples of this line of evolution would include plastic properties that change with conditions in a favorable way and motor oils whose viscosity changes in a controlled way with temperature.

Recent inventions that illustrate this principle are pulsed micro jets for needle free injections, flexible piano keyboards, license plate holders that change their framing announcement, windshield wipers whose speed varies with that of the car, and radio noise levels in cars that vary with the speed of the car.

PepsiCo is experimenting with changing store stocking and pricing with the known paycheck cycles in the local community.

4. Field Progression

All of us can remember the progression of how communication has evolved. Ages ago, we sent messages with a mechanical field ("tom tom" drums). Then we used a thermal field (smoke signals), whose image could be seen at a far greater distance. We then used a chemical field (ink) to write letters whose reach was thousands of miles. Then Alexander Graham Bell came along with an electronic field (the telephone). We now use an electromagnetic field (wireless) in the form of cell phones. And how many things can your cell phone do that used to require separate devices that illustrate the system integration principle previously mentioned. This same field progression can be seen in how we make presentations. Stick pointers were replaced with laser pointers and, in the middle; we saw dynamism (previous line) in the form of a flexible metal pointer.

Where have we seen this progression in the chemical industry? Our process control used to be manual operated manual valves adjusted by operators based on visual observations. Thermocouples (a thermal field) were introduced that allowed this decision to made with actual information. On line chemical analysis now allows actual compositional feedback. Pneumatic control (a dynamic mechanical field) made this task easier. Chemical and electronic fields in the form of valve actuators and on line analysis took us to another level of closed loop control. With electromagnetic fields, we now have data transfer wirelessly. And we should not forget about this line progression in terms of cooking---thermal ovens to microwaves.

In chemical mixing, we have moved down this line with mechanical mixing to surfactants to ultrasonics. In specialty and heat sensitive drying, we have moved to microwave drying vs. a thermal field. In chemical polymerizations, we have moved for virtually uncontrolled thermal polymerizations to highly specific chemical catalysts and we now see some visions of microwave catalysts.

We now see UV sterilization in place of chemical sterilization in some instances

5. Examples and Proactive Use of Trends

Since these lines of evolution are predictive, how can we use them in a pro-active way instead of waiting until we see a competitor's activities and react to it?

A simple way of doing this is to use a simple graphical analysis based on the simple game of "tic-tac-doe":

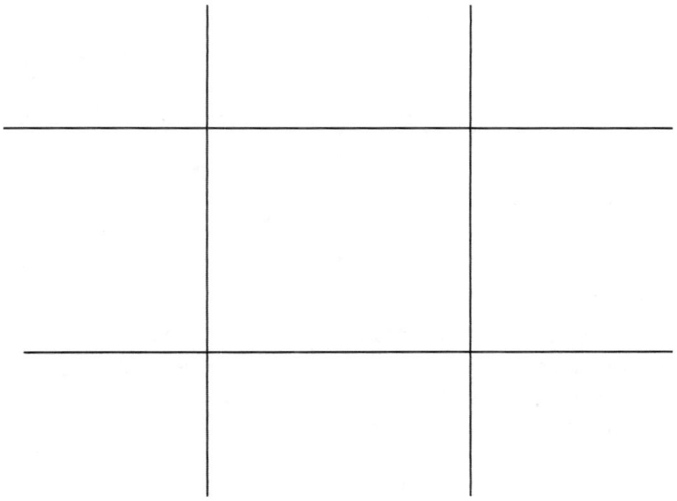

Instead of using this as a random game, let's use it as a framework for a discussion of the aforementioned principles:

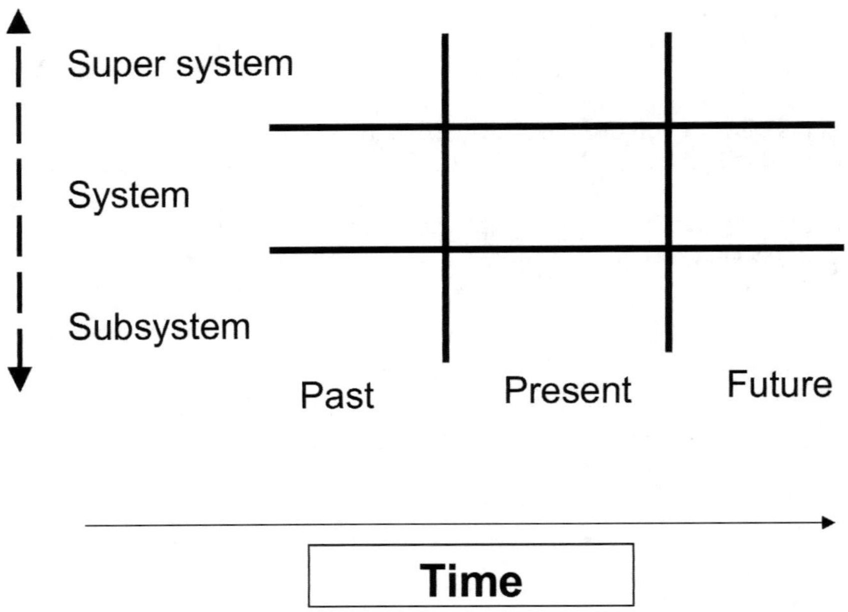

In the development and evolution of technical systems, progress is inevitably upward and to the right, i.e. to the upper right hand corner. Let's review several of the previous examples to illustrate. If we started with the analysis of "making copies", the diagram would look like this:

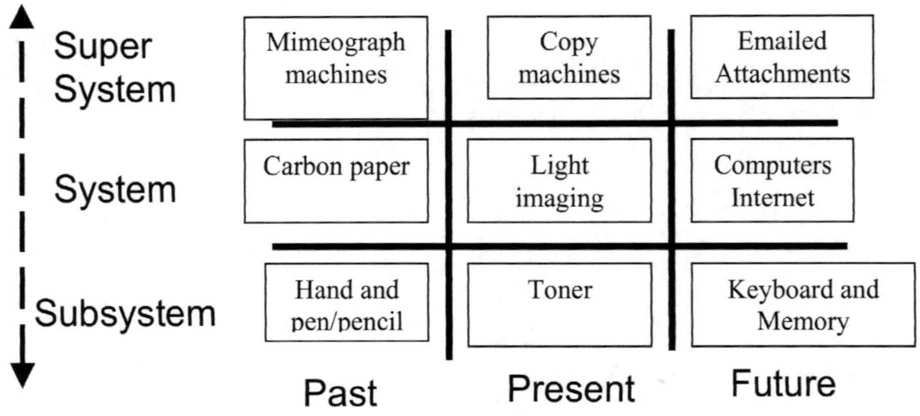

If we look left to right, we see the progression from mechanical to electronic fields. If we look from bottom to top, we see the system integration trend. If we had been the head of R&D at a company in the typewriter business and were aware of these trends we would have taken the lead in developing a system that allowed copies to be transferred without the use of a traditional keyboard.

Let's look at one of these diagrams for a consumer need: cleaning the shower.

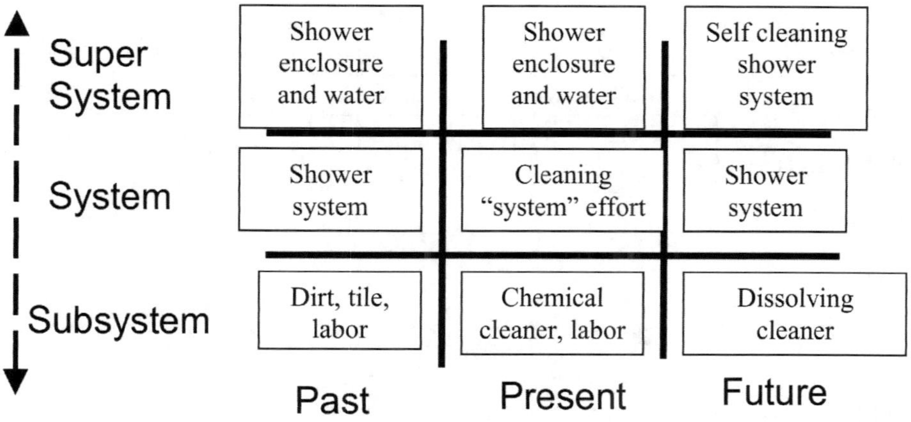

Planning for the future via TRIZ thinking would involve thinking about getting to the upper right hand corner while advancing in field use and integration into the super-system. S.C. Johnson's new self cleaning shower system comes close where a refillable cleanser is simply hung around the shower head and the chemical and hydraulic energy replaces the mechanical energy involving labor previously used.

Let's take a look the chemical business of "bleaching paper". Thousands of tons of chlorine are sold into this market every year.

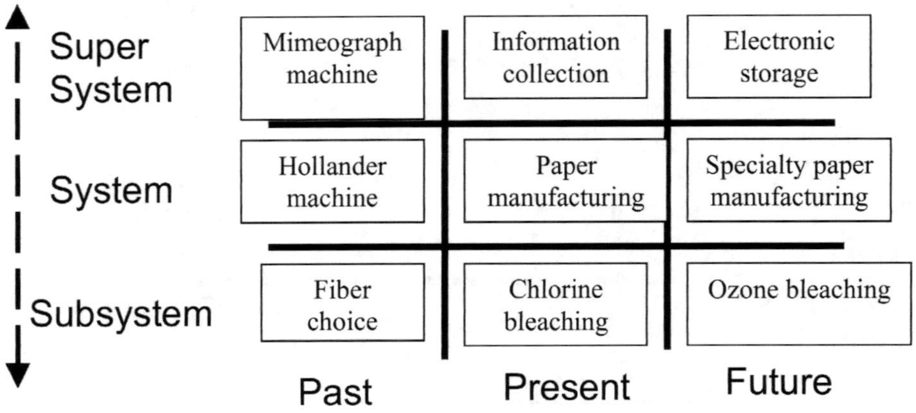

Each of us has seen the trend toward electronic document storage which requires no paper or bleach. We have also seen, in the past year, the transition of previously mailed "newsletters" and some major print magazines (US News and World Report) and newspapers (Ann Arbor News) to a totally electronic format, virtually eliminating the use of paper. Chorine is no longer needed because the paper is not needed. If the R&D department of a chlorine chemical supplier is basing their R&D program on how to better optimize the use of chlorine in a paper mill, they will be surprised over time as business drops off. Even looking for superior "bleaching" approaches that make paper brighter is a waste of long term resources. Would investing in Amazon's new of site storage business be a better investment? It's a key long term strategic decision—are you in the bright paper business or in the business of providing the optimum way of storing and recording information? Very different core competencies are required.

This same example shows the interplay with the dynamism line. Electronic storage provides ways of instantly reorganizing and instantly deleting information. Paper does not have that flexibility.

We can take a look at one other materials example. The chemical industry is a major supplier of rubber and elastomer materials to the tire industry. We spend a great deal of money on working with tire suppliers to optimize elastomer formulations, adhesion to tire cords, provide resistance to blow outs, etc. Let's take a look at this via this method:

	Past	Present	Future
Super System	Wheel assembly	Wheel and tire assembly	Wheel without tire
System	"Tire" with spoked wheel.	Automobile Tire	Automated tire response
Subsystem	Natural rubber	Elastomers	Specialty elastomers w/blow out prevention

If we look at the up0per right hand box, we see that what TRIZ predicts will happen is that the function of the tire will be performed without existing. An example is shown in the new Michelin "Tweel"™:

What are the implications for the material supplier? What is the new volume of elastomer needed? Are its properties the same? What new problems or opportunities might be available?

6. Summary

The lines of evolution contained within the TRIZ methodology are powerful strategic thinking tools that can assist us in formulating R&D projects, analyzing our business, and assisting in deciding who to talk to with regard to long term strategic planning.

7. Resources

Websites:

www.innovation-triz.com
www.aitriz.org
www.triz-journal.com
www.etria.net
AIChE/ASME course in TRIZ: Sept. 28-30, Las Vegas and Dec.7-9, Atlanta
http://catalog.asme.org/Education/ShortCourse/TRIZ_Theory_Inventive_Problem.cfm

Management Division: Managing Innovation

Different techniques for managing innovation are better suited in different environments, and the tools used in one will not necessarily be helpful in another. The strategy for protecting innovation will also change depending on the size of the company and the area of technology. The presentation will highlight different strategies of innovation protection. This presentation will then focus on different techniques used in companies ranging from the corporate giant through the entrepreneurial start-up, why those techniques are best suited for that particular environment and identify best practices for marketing, legal and research and development departments working together. The presentation will also discuss different products on the market which assist companies in managing innovation, weighing new ideas and analyzing patents.

Authors: Phil McCann, Kimberly Gatling, Stan Antolin, Clara Cottrell and Dave Krasnow
Presenter: Dave Krasnow
Company: Smith Moore Leatherwood, LLP

Corporate and Business Transactions Group

Smith Moore Leatherwood's Corporate and Business Transactions group have built relationships that have endured for decades, acting as an outside general counsel, strategic business advisor, and one-stop legal resource for businesses of all types and sizes. We take a proactive approach in helping our clients solve day-to-day problems and developing strategies for future growth or dispositions.

We know the fundamental measure of a good client relationship is how little our clients worry about legal matters and how much they focus on the success of their business and their patent portfolio. Our business and Intellectual Property lawyers act as a primary contact, handling day-to-day legal matters and referring clients to other Smith Moore Leatherwood lawyers with skills in particular disciplines as needed.

Our attorneys represent a variety of business concerns, some focusing on distinct industries such as health care, education, textiles and polymer chemistry. We have hands-on experience dealing with all forms of business enterprises, including limited liability companies, partnerships and professional corporations. The group offers advice on start-up issues and regulatory compliance, and regularly assists with contract issues such as employment contracts, licensing agreements, supply contracts, consulting agreements, purchase and sale forms, and warranty agreements. We are skilled in handling mergers, acquisitions and other transactions involving business combinations, representing buyers, sellers and equity investors.

SMITH MOORE LEATHERWOOD

David M. Krasnow

Two Hannover Square
Suite 2800
434 Fayetteville Street
Raleigh, North Carolina 27601

T (919) 755-8741
F (919) 838-3171
dave.krasnow@smithmoorelaw.com

Dave Krasnow practices in the field of chemical patents and has more than 20 years of patent experience handling complex patent matters in fields including polymer chemistry, inorganic chemistry, optics, ocular science, pharmaceuticals, industrial gases, raw and specialty chemicals, coatings, refractory materials, material deposition and medical devices. Mr. Krasnow's practice encompasses counseling clients in all patent matters and rendering opinions on complex matters including invalidity and infringement. Mr. Krasnow often conducts due diligence on all Intellectual Property-related matters for corporate clients involved in complex merger, acquisition, divestiture and IPO activities, where IP rights and valuation play a central role.

Mr. Krasnow has served as in-house counsel with Bausch & Lomb, Inc. (Rochester, NY), and has practiced with private firms in Rochester, NY and Washington DC. Prior to entering law school, Mr. Krasnow served as an Assistant Clerk of the United States Supreme Court in Washington, DC. Mr. Krasnow lectures frequently on various Intellectual Property topics, and has taught Advanced Patent Law courses at the University of North Carolina Law School and lectures annually at Duke University's Fuqua School of Business.

SMITH MOORE LEATHERWOOD.

Stanislav Antolin

300 N. Greene Street
Suite 1400
Greensboro, North Carolina 27401

T (336) 378-5516
F (336) 378-5400
stan.antolin@smithmoorelaw.com

Stan Antolin is a member of the firm's Intellectual Property group. His experience ranges from counseling individual inventors through Fortune 100 companies. Stan's degrees in and experience with materials science and engineering processes have served this diverse group of domestic and international clients well as they have sought to establish a competitive advantage through the use of intellectual property. Examples of business areas in which Stan's clients participate include tooling, medical device, pharmaceutical, biochemical, materials technology, encryption technology, data management, software and specialty chemical. He is involved in the preparation and prosecution of patent applications and appeals before the Board of Patent Appeals and Interferences; preparation of patentability, validity, freedom-to-use and noninfringement opinions; intellectual property licensing, and litigation support.

Most recently, Stan served as an associate with an intellectual property boutique. Prior to that position, he served as patent counsel for Kennametal Inc., in Latrobe, Pennsylvania. Stan started his career in intellectual property with a five-year stint as a patent agency/engineer for Lanxide Corp. in Newark, Delaware, where he had previously worked as a research engineer. Stan has prepared and prosecuted numerous applications for patent protection in the United States, Japan, and throughout Europe. He is also the co-inventor on six U.S. patents.

Publications

"Patent Law within the Context of Intellectual Property Law" and Appendix J: "Comparisons of Types of Intellectual Property" in Patent Settlements, Negotiations, & Litigation (Aspatore Books), pp. 87-101 & pp. 193-195 (2006)

"Formation of Al_2O_3/Metal Composites by the Directed Oxidation of Molten Al-Mg-Si Alloys: Part 1, Microstructural Development," with A.S. Nagelberg and D.K. Creder, Journal of the American Ceramic Society, Vol. 75, No. 2, pp. 447-453 (1992)

"A Study of the Reaction of CO on Liquid Iron Alloys," with R.J. Fruehan, Metallurgical Transactions B, Vol.18B, pp. 415-420 (1987)

SMITH MOORE LEATHERWOOD.

Kimberly Bullock Gatling

300 N. Greene Street
Suite 1400
Greensboro, North Carolina 27401

T (336) 378-5356
F (336) 378-5400
kim.gatling@smithmoorelaw.com

Kim Gatling concentrates her practice in patent and trademark prosecution, licensing, and litigation. Particularly, she prosecutes computer software, business method, and mechanical patent applications, as well as trademark applications, before the United States Patent and Trademark Office. She also drafts and negotiates agreements for IT transactions, including software development and maintenance agreements, and advises clients on computer law and Internet issues. Ms. Gatling further advises clients with regards to copyright registration and litigation.

Publications

"Claiming Your Biotechnology Company's Intellectual Property: Guidelines for Claim Construction of United States Patents" – National Law Journal, Vol. 25, No.17, December 16, 2002

"Safe Share – What Colleges and Universities Can Do to Stop Illegal File Sharing" – Campus Counsel, Winter 2004

Phil McCann

300 N. Greene Street
Suite 1400
Greensboro, North Carolina 27401

T (336) 378-5302
F (336) 433-7437
phil.mccann@smithmoorelaw.com

Phil McCann's practice focuses on intellectual property and in particular patent law. He prosecutes patent applications before the United States Patent and Trademark office and regularly advises clients on patent infringement and patent licensing issues as well as management of patent portfolios.

Prior to joining Smith Moore Leatherwood, Mr. McCann was Chief IP Counsel for Great Lakes Chemical Corporation in Indianapolis, after having served as a patent attorney at Clariant and Hoechst Celanese in Charlotte for 10 years.

Publications

"Guidelines for Claim Construction of U.S. Patents" – BT Catalyst, February/March 2003

"Claiming Your Biotechnology Company's Intellectual Property: Guidelines for Claim Construction of United States Patents" – National Law Journal, Vol. 25, No.17, December 16, 2002

Clara R. Cottrell

300 N. Greene Street
Suite 1400
Greensboro, North Carolina 27401

T (336) 378-5597
F (336) 433-7576
clara.cottrell@smithmoorelaw.com

Clara Cottrell is a member of the firm's Corporate and Intellectual Property group. Ms. Cottrell's practice is related to polymer chemistry, biomedical devices, general mechanical and chemical advances, trademarks and copyrights as well as business transactions and contracts. Prior to joining the firm, Clara clerked for the Honorable Ben F. Tennille of the North Carolina Business Court in Greensboro, NC.

Clara graduated from Wake Forest University School of Law in 2007. While there, she was active in the Intellectual Property Law Association and Women in Law. Clara continues to volunteer her time to Wake Forest as a mentor and speaker on various panels. Clara is also active in the ABA Section of Science of Technology. She served as the Law Student Liaison while in law school and is currently a columnist for *The SciTech Lawyer.*

Publications

"Most Valuable Patent: The Use of Natural Phenomena in Patents" 7 Wake Forest Intellectual Property Law Journal 251 (2007).

"The Supreme Court Brings A Sea Change With KSR International Co. V. Teleflex, Inc." 42 Wake Forest Law Review 595 (2007).

Managing Innovation to Maximize Patentability

Neal Seth, Baker Hostetler LLP, Washington, DC

This presentation will focus on guiding principles and best practices to manage innovation so as to best leverage patentability.

See more of this Session: Managing Innovation
See more of this Group/Topical: Management Division

Silence Fails

The Five Crucial Conversations for Flawless Execution

CONTENTS

AN INTRODUCTION TO *SILENCE FAILS*

The Five Crucial Conversations for Flawless Execution

Upwards of 91 percent of all large-scale corporate initiatives fail—initiatives that have been designed to save a company from impending disaster, increase organizational competitiveness, or just accomplish strategic objectives. They consume unplanned resources and miss schedules, they leave behind the stigma of failure, and they cost leaders careers.

Yet many organizations continue along the same precarious path year after year—accommodating behaviors that contribute to these costly mistakes.

How can we avoid repeating history? Why do projects that demand so much of a company's top leaders' time and attention collapse? Do people just not know how to plan? Do they fail to involve experts? Are they hopelessly incapable of making accurate projections? No. History repeats itself when teams and organizations—made up of people—fail to speak up well and be heard. Unfortunately, a deadly form of corporate silence lies at the root of all failed projects. Like it or not, project problems are people problems.

In the following pages, you will find some startling results that will change the way you look at your next initiative, program, or project—and demonstrate that you can get better. This groundbreaking research, *Silence Fails*, outlines findings that are simultaneously daunting and hopeful. For the first time, *Silence Fails* accurately contrasts the causes and costs of failed initiatives with the impressive returns of investing in skills that can help anyone navigate the pitfalls of project execution.

How Can You Improve?

For more than twenty-five years, the founders of VitalSmarts have researched methods for bringing about systematic and lasting change. Based on this ongoing research, VitalSmarts has helped thousands of organizations—including more than three hundred of the Fortune 500—realize quick, hard-hitting results.

The following are some tips for getting started:

1. Take the *Silence Fails* Assessment found at the end of this report. This assessment highlights the points at which your initiative, program, or project is likely to fail.

2. Distribute this report to help others become aware of the problems it reveals.

3. Once you've increased awareness and appreciation for the issues, it's time to move to action. A VitalSmarts Representative can help you develop an action plan that will meet your specific needs. Call today to find out how you can begin to change your organization.

Other tips for change are available in the Conclusion section of this report and at www.silencefails.com.

VitalSmarts™

THE *SILENCE FAILS* STUDY

The Five Crucial Conversations for Flawless Execution

The Silent Crisis

This year, leaders across the world will make an awful bet. They will bet a substantial portion of their organizations' budgets on ventures that have less chance of paying out than a red or black bet on a roulette wheel. And yet, as in every year for many decades, leaders will continue to put the money down.

Their preferred hedge against failure? Formal systems. Over the past twenty years, project professionals and management experts have focused on improving the formal systems related to program governance, project management, and project-related technologies.

While these new approaches have substantially improved results, today two-thirds or more of initiatives, programs, or projects still miss their mark. Something is clearly missing. Beginning in late 2005, VitalSmarts and The Concours Group partnered to dive below the formal systems to identify what is missing when project execution fails.

This study, *Silence Fails: The Five Crucial Conversations for Flawless Execution,* demonstrates that project leaders can substantially improve their organization's ability to execute on high-stakes projects and initiatives by breaking a code of silence on five astoundingly common yet largely undiscussed and ignored problems that contribute significantly to almost all project failures.

KEY FINDINGS

Breaking a code of silence on five astoundingly common yet largely undiscussed and ignored problems contributes significantly to project success.

Situation Analysis

Senior executives frequently bet their companies on high-stakes efforts like major product releases, strategic IT projects, organizational restructurings, fast-paced downsizings, or aggressive quality initiatives. And these bets rarely pay off as anticipated. With estimated failure rates ranging from 72 to 91 percent,[1] companies' collective inability to execute on major projects costs hundreds of billions of dollars a year. For example, it's estimated that of the $255 billion spent per year on IT projects in the U.S., more than a quarter is burned up in failures and cost overruns.[2]

The failure of the cutting-edge baggage handling system at the Denver International Airport provides a dramatic example of this inability to execute. With twenty-six miles of conveyors, the system promised to carry bags more quickly and reliably from passengers to planes and back again. Instead, the breakthrough concept racked up hundreds of millions of dollars in cost overruns, delayed the airport's opening by more than a year, and drained ten years of additional resources in various salvage attempts before leaders admitted failure.[3]

In addition to sapping organizational performance, these project failures cost careers. Now more than ever, CEOs are under pressure to either get results or get lost. In 2005, CEO turnover doubled from the year before.[4] In just the past five years, close to two-thirds of all major companies have replaced their CEOs. CIOs are similarly vulnerable, with a quarter losing their jobs each year.[5] Studies suggest that an inability to deliver on critical projects is a primary reason for this alarming rate of dismissals. Their shortcomings were less about strategy and decision making than about their ability to execute these plans.[6]

KEY FINDINGS

When these problems are not openly and skillfully discussed, projects of all sizes fail miserably.

Five Crucial Conversations

Silence Fails suggests that leaders can significantly increase their organization's ability to execute on high-stakes projects and initiatives by focusing on the five crucial conversations outlined below. The study also found that increased effectiveness at any one of the five crucial conversations that address these problems improves project performance by 50 percent to 70 percent or more.

The five crucial conversations identified by the study are the most prevalent and most costly barriers to project success. Each brings with it myriad misses and overruns that plague projects, teams, and organizations. *Silence Fails* addresses the cost, culture, and long-term dynamics of each. They are:

1. **Fact-Free Planning.** A project is set up to fail when deadlines or resource limits are set with no consideration for reality.
2. **AWOL Sponsors.** A sponsor doesn't provide leadership, political clout, time, or energy to see a project through to completion.
3. **Skirting.** People work around the priority-setting process.
4. **Project Chicken.** Team leaders and members don't admit when there are problems with a project but instead wait for someone else to speak up.
5. **Team Failures.** Team members are unwilling or unable to support the project.

This study confirms that some projects are plagued by more of these problems than others—likely due to high levels of interdependence among levels and functions. And when these problems are not openly and skillfully discussed, projects of all sizes fail miserably. Those who effectively discuss these issues have a positive impact on the objectives of their projects.

Key Findings

Silence Fails demonstrates that these five problems are not a project death sentence, but they significantly impact the results a team or organization may be trying to achieve. Findings include:

- 90 percent of project leaders regularly experience one or more of these problems with nearly 40 percent of their projects.

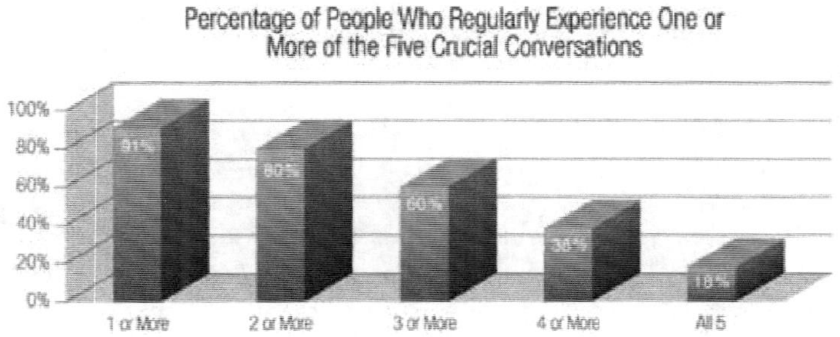

- 91 percent report that these problems are difficult if not impossible to resolve in their current organization.
- 61 percent attempt to speak up about the issue, yet only 14 percent feel they speak up in a skillful way that has a lasting effect on the problem.
- Up to 88 percent claim that these issues continue over the life of their project and beyond.
- Between 82 percent and 87 percent of projects report significant negative impact when they fail to address budget, scope, and schedule issues.
- Team morale suffered significant setbacks among 73 percent of projects studied. 43 percent of projects continued to suffer from a long list of problems even after the project ended.

Silence Fails demonstrates that whether these five problems can be discussed determines whether or not they will derail an initiative. The data show that when people discuss these problems in effective ways, projects come back on course. And when one or more of these five problems is not confronted—or not confronted well—it festers, sets off workarounds, and produces politics. In the end, project failure is almost inevitable.

KEY FINDINGS

When one or more of these problems is not confronted —or not confronted well —it festers, sets off workarounds, and produces politics.

The Good News: Speaking Up Makes a Difference

While the data show that fewer than one in five project leaders effectively engage in the crucial conversations needed to solve these five problems, the good news is that one in five *does*. Much can be learned by studying their successes and whether these skilled individuals produce different results than their peers. Research on this contingent of crucial conversations successes helps answer questions like:

- Does it affect project success when project leaders speak up effectively?
- Can others be taught to speak up more skillfully with similar results?

The resounding answer to each is "Yes." In each of the five key problem areas, the study found a clear difference between speaking up and speaking up well. Across the problem areas, about half of leaders make some attempt to speak up. But most are ineffective. Some speak up but water down their concerns, so the issues are never fully aired. Some speak up but do so in a way that provokes defensiveness from others. And a handful—about one in eight—are able to share their full concerns and, by the end of the conversation, feel their views are understood.

KEY FINDINGS

Unless and until leaders take measures to ensure their environment is conducive to holding crucial conversations, a significant number of these issues will remain unaddressed, invisible, and fatal.

The research also showed that while the skill of the initiator is a key ingredient in ensuring these crucial conversations are held well, the receptiveness of the other party is similarly important. Project leaders who are adept at holding crucial conversations are far more effective in dealing with prickly and defensive executives. But their skill doesn't guarantee their concerns will be addressed. Unless and until leaders take extraordinary measures to ensure their environment is conducive to holding crucial conversations, a significant number of issues will remain unaddressed, invisible, and fatal.

However, when a skillful leader creates even a moderately safe environment:

- The likelihood of a project failing to meet budget is reduced by almost half.
- When project leaders effectively step up to these crucial conversations, hitting schedule is at least 40 percent more likely, quality improves almost 60 percent, and the likelihood a project will end with strong morale and intact stakeholder relationships is almost 70 percent greater.

Next Steps for Senior Leaders

Although *Silence Fails* focuses on important findings that can predict and explain failure, the most important implication of the study is the potential leaders have to influence success.

Senior leaders can predict and prevent the failure of high-stakes business initiatives by creating a culture where the five conversations are held quickly and effectively. *Silence Fails* also provides insights and recommendations on how senior leaders can develop a business case for change, measure behaviors, invest in skills, hold senior management accountable, and make heroes of early adopters.

Organization of This Report

This report contains the results of the *Silence Fails* study. Each of the five crucial conversations for flawless execution is explored, including the following:

- A composite case study, "The Last Roundup," that describes the issues the *Silence Fails* research team documented in actual projects.
- Key findings as they relate to each specific crucial conversation—including the extent of each problem, the cost of not speaking up, and the effect of speaking up.
- Actual comments from study participants that further reflect the environment and effects of these crucial conversations when they go unaddressed.
- A conclusion that identifies best practices for leaders who want to help their organizations improve their ability to step up to and hold these five crucial conversations to improve project success.

Section Notes

1. "CHAOS Chronicles," Standish Group, 2004, and "The Strategy Focused Organization," Kaplan and Norton, 2001.
2. "CHAOS Chronicles," Standish Group, 2004.
3. "Dependable Software by Design," *Scientific American*, June 2006, 69.
4. Challenger, Grey, and Christmas, cited in *USA Today*, 10 April 2006, 2B.
5. "Work Management. Why Can't Information Managers Manage?" John J. Lucas, PMI Seminar/Symposium. Proceedings, 1995, 304–310.
6. Study by LeadershipIQ.com.
7. Based on over twenty five years of research, VitalSmarts continues to conduct field studies that demonstrate significant, sustainable results brought about through organizational change initiatives. These efforts focus on behavior modification as opposed to system design and strategy development. See www.vitalsmarts.com for case studies and additional information.

CASE STUDY:
THE LAST ROUNDUP

A project leader in a major business services company, Bob Kelley led an effort codenamed "Roundup" to insource the company's billing system. The history of the billing system was complex:

- The billing system was the lifeblood of the company—responsible for drawing hundreds of millions of dollars into the company on a monthly basis by issuing timely and accurate bills to more than 20 million customers.
- For twenty years the billing system had been handled by a vendor.
- For three years Kelley led a sometimes unruly and loose confederation of more than 1,000 professionals drawn from various functions in his company, as well as contractors, to change over to a company-owned system. Roundup was a particularly tough assignment given that Kelley was the line manager of only a few of these people, and his mission depended on the unflagging support of *all* of them.

The stakes couldn't have been higher. And Roundup's status could not have been worse. By the time we met Kelley, the project was already 100 percent overdue and 300 percent over the original budget—an investment of well over $250 million to that point with no end in sight. Kelley believed, however, there was a light at the end of the tunnel. The first cutover date for a small contingent of customers was two weeks away. As he walked out of a "release review," he believed all was in order and the chance of success was high.

One month later, Kelley was handed his head. The release was a disaster. The vendor managing the legacy system failed to transfer data when required. Once they did, the new bills were plagued with errors, and the system suffered from multiple "Severity 1" glitches—causing it to crash uncontrollably during the billing cycle. When he was dismissed, Kelley was escorted to his desk by a security guard who then walked him out of the building.

Diagnosis

Kelley's demise was the direct result of failure to effectively address the five issues identified in *Silence Fails.* Each of the five key problems emerged at one time or another during the course of the initiative.

Conclusion

The complete undoing of Roundup was not guaranteed by the existence of these problems but by how they were handled. Most were never fully discussed with the people who needed to hear them. Others were raised but then muffled by the very executives who should have listened and taken action.

ABOUT THIS CASE

This case follows a composite character, Bob Kelley, through a major organizationwide initiative, and describes issues the *Silence Fails* research team documented in actual projects.

Throughout the remainder of this study, pertinent *Silence Fails* findings will be applied to the case study.

VitalSmarts™

PROBLEM 1

Fact-free planning reflects bad planning behaviors at every level.

Extent and Costs

Of the 85% of study participants who reported the prevalence of fact-free planning in their projects:

- 39% of their projects were affected
- 91% said the issue was difficult or impossible to confront or resolve
- 61% spoke up at all
- 14% spoke up skillfully and had some effect on the problem
- 19% were able to solve the problem

88% reported that the problem persisted,* causing operational impact as follows:

- 82% of projects exceeded their budget
- 82% missed deadlines
- 82% had missing or wrong functionality or quality problems
- Team morale was damaged on 73% of projects
- 43% had a long list of problems to be resolved after the project ended

* Continued for the life of the project and beyond.

CRUCIAL PROBLEM #1:

FACT-FREE PLANNING

A project is set up to fail when deadlines or resource limits are set with no consideration for reality.

Fact-free planning describes ways leaders set project parameters without involving—or by falsely involving—project leaders. Project leaders are often given sets of deliverables, budgets, and timelines with no input or opportunity to determine what the project will really take. At other times they are asked to calculate the resources and time needed to achieve a scope of work, but then their estimates are ignored. In either case, the result is a set of timelines, budgets, and deliverables that are unrealistic from the beginning.

Very often, unrealistic constraints get placed on projects. An executive or customer will make commitments to another stakeholder without the project team's consent and will then present the finished plan as a done deal. Over time, teams that are subjected to these mindless demands begin to pad budgets and timelines—anticipating that their estimates will be ignored.

Fact-free planning reflects bad planning behaviors at every level. When project leaders realize these practices are taking place, they must be willing and able to call the bluff. If they avoid this crucial conversation and either commit to something they know can't happen or fake their way to success, they set themselves and their projects up to fail.

Similarly, executives who avoid discussing their doubts about an estimate's validity or the team's competence instead use their power in a way that generates *political* rather than *valid* agreements. Then when failures follow, their doubts about the team are confirmed, and they feel justified in using more fact-free planning to re-establish their sense of control.

The only way out of this vicious cycle is for project leaders and executives to candidly and effectively express their suspicions, doubts, and data. While these crucial conversations aren't easy, they are the only path to rational commitments.

How common are these problems? Eighty-five percent of project leaders are faced with "fact-free planning." Key drivers include:

Indicators of Fact-Free Planning

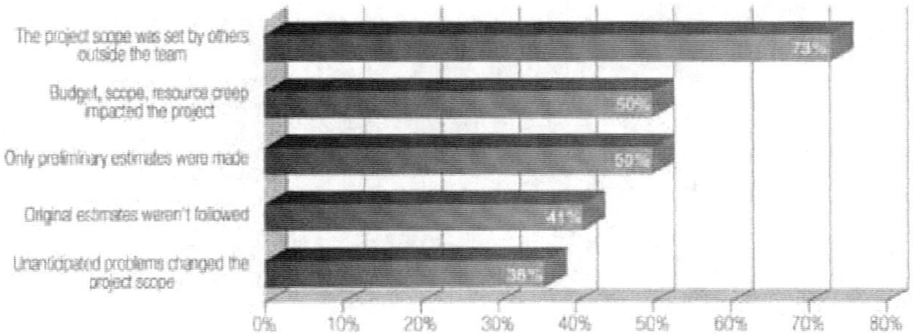

Just how this plays out with project teams is better illustrated by these key findings:

- When project leaders face these problems, they know the stakes are high, and 61 percent make some effort to raise their concerns—often in an understated way.
- More than nine out of ten project leaders describe this type of conversation as between "difficult" and "impossible" to resolve.
- In fact, only 19 percent are able have the conversation in a way that gets results. When project leaders are unable to succeed at this conversation, their projects turn into slow-motion train wrecks.

When the conversation fails, the impending project failure becomes widely anticipated.

- Nine out of ten project leaders share their concerns with project team members and with others across the organization.
- Team morale takes a hit in nearly three-quarters of these cases as people become aware that they're headed toward failure. At this point in the project, the unrealistic specs are rarely resolved; 88 percent of the time they remain for the life of the initiative or beyond.
- 9 percent of these projects come in more than $1 million over budget.
- 70 percent of the time, people take inappropriate shortcuts, cut documentation, or skip testing in order to meet the specifications, and in the end these projects fail.
- In fact, 61 percent achieve the hat trick of project failure; they miss their deadlines, go over budget, *and* fail to achieve their deliverables.
- 43 percent end up with a long list of problems to resolve after the project itself has ended.

Participant Comments

"Here's how it works here. Year 1: Six Sigma is the hot thing and you get all the resources you need. Year 2: They put the squeeze on you—arbitrarily cutting by X percent. Year 3: You're not meeting goals because you have no resources, and they kill it." – *Manager, pharmaceutical firm*

"We ask you to do a project. You come back and say, 'I need twenty-five resources and $2.5 million.' And your sponsor says, 'Okay, how about three resources, clock time, and no money?' and we say 'Sounds right.' You have to, because you can't challenge." – *Department Head, food services firm*

"There's no such thing as planning for a project. There's the real date and the date you're told to do it. Part of the front end of a project should be identifying and refining your scope. What is realistic? In previous experience it's not about what is the true scope and how many resources and how long. Senior manager A spoke with senior manager B—we're going to have this done by the end of February. We set an end date and work backward. So we cut out testing and other quality processes." – *Director, financial services firm*

Impact of Speaking Up

The 19% who addressed this problem** improved quality and functionality by 69% while reducing:

- Cost overruns by 29%
- Schedule delays by 17%
- Inappropriate shortcuts by 47%
- Persistent problems* by 58%

* Continued for the life of the project and beyond.
** Compared to those who didn't address the issue (these differences are all statistically significant, $p < .05$)

Application
THE ROUNDUP CASE

When Kelley took over Roundup, he was given one year to complete the insourcing, and a budget of $45 million. Ideally, this schedule and budget would have been the result of a careful planning process that involved Kelley, but that's not the way it happened. Instead, Kelley's executive leaders were playing a common game in project execution—they fished for the best deal they could get. They set a stretch goal and hoped the performance pressure it created would motivate him to achieve it. Kelley was a good soldier and tried to make it work, but he knew from the beginning that it couldn't.

PROBLEM 2

When the project sponsor is AWOL, the project team is stranded and exposed.

Extent and Costs

Of the 65% of study participants who reported the prevalence of AWOL sponsors in their projects:

- 33% of their total projects were affected
- 88% said the issue was difficult or impossible to confront or resolve
- 47% spoke up at all
- 11% spoke up skillfully and had some effect on the problem
- 17% were able to solve the problem

75% reported that the problem persisted,* causing operational impact as follows:

- 75% of projects exceeded their initial budget
- 85% missed deadlines
- 74% had missing or wrong functionality or quality problems
- Team morale was damaged on 69% of projects
- 31% had a list of problems to be resolved after the project ended

* Continued for the life of the project and beyond.

CRUCIAL PROBLEM #2:

AWOL SPONSORS

A sponsor doesn't provide leadership, political clout, time, or energy to see a project through to completion.

Project sponsors are responsible to provide leadership and political support. And they frequently don't. When the sponsor is AWOL, the project team is stranded and exposed. They're sent off to accomplish a task and don't have the firepower needed to implement the project. For example, key leaders whose help or resources are needed to enable the project may not come through, and the sponsor who has the organizational muscle to hold them accountable fails to do so.

In many cases the sponsor who should have fought the political battles (returning calls, following up with commitments, and so on) ultimately leaves project leaders to fend for themselves. Rather than directly address the issue, they begin working around it.

Key findings include:

- 65 percent of project leaders experience problems with AWOL sponsors.
- About half of those make some attempt to bring up their concerns.
- 88 percent of study participants indicate that conversations on this subject are difficult if not impossible to have in their organizations.
- Fewer than one in five project leaders who suffer from AWOL sponsors are able to hold the crucial conversation with their sponsor in a way that solves the problem.

When the project leader fails to resolve AWOL sponsor problems, projects are destined to fail—and the costs add up.

- More than three-quarters come in substantially over budget, behind schedule, and/or below specifications.
- In fact, 11 percent come in more than a million dollars over budget.
- Team morale is damaged more than two-thirds of the time.

Participant Comments

"My project from the president was, 'We need to eliminate $50 million in cost by moving three major manufacturing programs from North America to our Asia operations.'

"My 'sponsor' was the VP of North America. I met with the North America person and he basically said, 'Over my dead body.' And this is my sponsor! I went to the president and said, 'What is up with this? How am I going to manage through this?' So he sent an e-mail to the sponsor and ordered him to help me. My sponsor is now my sworn enemy."
– Director, manufacturing firm

"I was working with people I've never met in another country and basically I was trying to drive something that the CIO wanted. I tried to get people to understand that this was what we were being told to do—that it was a directive, not a choice. I finally had to talk to the CIO and explain to him, 'I need help, you need to explain why we need to do this. You need to get people's buy-in and support.' I was afraid an e-mail would go out like 'This is what you need to do and you will do it NOW.' I don't think that would have helped the situation. Instead, he kind of backed off and the project died. But not for another year."
– Program Manager, pharmaceutical firm

Impact of Speaking Up

The 17% who were able to address this problem** improved quality and functionality by 50% while reducing:

- Cost overruns by 33%
- Schedule delays by 33%
- Persistent problems* by 70%
- Inappropriate shortcuts by 66%

* Continued for the life of the project and beyond.
** Compared to those who didn't address the issue (these differences are statistically significant, $p < .05$)

Application
THE ROUNDUP CASE

In Bob Kelley's case, the sponsor who should have fought the political battles required to get the old vendor to cooperate rarely returned calls, seldom followed up with commitments, and ultimately left Kelley to fend for himself.

When Kelley took on Roundup he made a fatal mistake. He was told through the grapevine that his sponsor was a bit of a deadbeat. Rather than directly address this issue, he began working around it. He called on other executives for help and enabled the problem by absolving his sponsor of responsibility.

PROBLEM 3

Skirting plagues most projects for their entire duration.

Extent and Costs

Of the 83% of study participants who reported the prevalence of skirting in their projects:

- 33% of their total projects were affected
- 79% said the issue was difficult or impossible to confront or resolve
- 51% spoke up at all
- 13% spoke up skillfully and had some effect on the problem
- 16% were able to solve the problem

81% reported that the problem persisted,* causing operational impact as follows:

- 78% of projects exceeded their budget
- 87% missed deadlines
- 80% had missing or wrong functionality or quality problems
- Team morale was damaged on 66% of projects
- 28% had a long list of problems to be resolved after the project ended

* Continued for the life of the project and beyond.

CRUCIAL PROBLEM #3:
SKIRTING

People work around the priority-setting process.

Powerful stakeholders and senior leaders often skirt the formal decision-making, planning, and prioritization processes. They need what they need, and they don't want to be burdened with practical considerations. So they work around the process. The results are often outrageous overcommitment, disappointment, and burnout. Projects get approved for which there are no resources, scope creep bloats approved projects far beyond the resources originally budgeted, and team members deliver a succession of disappointing results and suffer from battered morale.

Key findings include:

- 83 percent of project leaders contend with stakeholders and others who skirt the formal planning and execution system. The most common descriptions of skirting include:

The Many Faces of Skirting

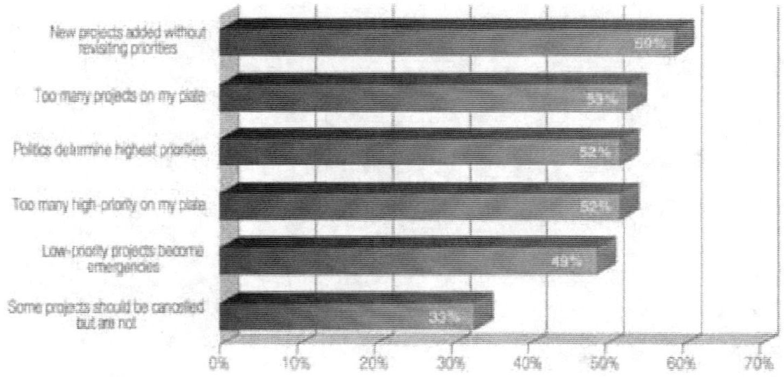

- 79 percent of project leaders describe the discussion as between "difficult" and "impossible" to resolve.
- Only 13 percent are able to skillfully raise their concerns in a way that gets heard.

With so few people able to succeed at this crucial conversation, it's no surprise that skirting plagues projects for their entire duration. Other key findings include:

- Four out of five of these projects suffer from skirting issues for the life of the project or beyond.
- 78 percent of these projects go over budget.
- 87 percent miss their deadlines.
- Two-thirds leave teams with battered morale.
- 80 percent fail to achieve their deliverables.

Participant Comments

"A real simple example: somebody might be walking with me to the coffee machine and they start talking about something. I have to stop and say, 'You realize this is just a conversation—that you did not just initiate a project. You cannot go back to your CIO three months from now and say, "We've been working with them for two months now." Talking to me on the way to the break room is not how you contract for our services.'" – *Senior Program Manager, food services firm*

"When a project manager says 'No' to a customer, here's what will happen: The customer will call the president of our company and say, 'You need to do this,' and he says 'Yes.'" – *Project Manager, food services firm*

Impact of Speaking Up

The 16% who were able to address this problem** improved quality and functionality by 44% while reducing:

- Cost overruns by 35%
- Schedule delays by 28%
- Persistent problems* by 71%
- Inappropriate shortcuts by 64%

* Continued for the life of the project and beyond.
** Compared to those who didn't address the issue (these differences are all statistically significant, $p < .05$)

Application

THE ROUNDUP CASE

The VP of Marketing in Kelley's company repeatedly injected additional requirements into the project with no consideration for their impacts on the budget or schedule. He did so through both direct intervention and deal-cutting with sponsors and executives. Kelley and his team were left to deal with the fallout. Once again—the problem was not only that the Marketing VP skirted decision-making processes—the problem was that no one confronted his violations in a way that stopped them.

Bob Kelley and his sponsors failed to step up to the crucial conversation when repeated skirting made the scope of his project impossible to fulfill. As a result, the violations continued and added momentum to the oncoming train wreck.

PROBLEM 4

When project participants play project chicken, the status and review process becomes a joke.

Extent and Costs

Of the 55% of study participants who reported the prevalence of project chicken:

- 31% of their total projects were affected
- 61% said the issue was difficult or impossible to confront or resolve
- 42% spoke up at all
- 13% spoke up skillfully and had some effect on the problem
- 26% were able to solve the problem

76% reported that the problem persisted,* causing operational impact as follows:

- 78% of projects exceeded their budget
- 86% missed deadlines
- 74% had missing or wrong functionality or quality problems
- Team morale was damaged on 54% of projects
- 23% had a long list of problems to be resolved after the project ended

* Continued for the life of the project and beyond

CRUCIAL PROBLEM #4:
PROJECT CHICKEN

Team leaders and members don't admit when there are problems with a project but instead wait for someone else to speak up first.

Once a project gets underway, it can derail when various subteams or team members fail to honestly report project risks. They engage in project chicken.

This costly game resembles the lunatic practice of driving cars head-on as a test of nerves to see who swerves out of the way first—or who is more "chicken." The corporate version is played when project leaders fail to admit they may fall short on deliverables and need more time. Instead, they hope some other group that has problems will speak up first. Whoever speaks up first will be blamed for causing the delay, but everyone who is behind will benefit.

When project leaders play project chicken, the status and review process becomes a joke. The team loses opportunities to gracefully respond to problems by revising goals, shifting resources, reorganizing plans, and more. Instead, the project hurtles forward on a collision course with failure while everyone watches—nervous but silent.

Almost every organization reported some form of project chicken, and more than half of project leaders say they face it regularly. Often, the enormous pressure to give good news sets project leaders up for failure by making it difficult to discuss risks. Such pressures also make it hard for team members to admit that a single problem could pose a risk to the entire project, even though the problems may be long-standing and complex. In this situation it's also hard to admit that shortcuts could pose a risk to the entire project. Project leaders often see these problems coming for months but fail to speak up in an effective way.

Key findings include:

- Roughly one in four project leaders who face this problem are able to effectively discuss project chicken with those who appear to be playing the game.
- When project leaders face these problems they know the stakes are high, but only 42 percent make some effort to raise their concerns.
- 13 percent of project leaders describe this type of conversation as "difficult" and "impossible" to resolve.

When the conversation fails, so do 76 percent of the projects.

- 86 percent of those miss their deadlines.
- 78 percent come in over budget.
- 74 percent are missing functionality or suffer from lower-than-desired quality.

Participant Comments

"There are plenty of cases where someone has known something was wrong but didn't speak up soon enough. So yeah, if you have project updates and you have good, good, good, good, good, then bad at the end of the day—then it's too late to do anything."
– *Project Manager, construction firm*

"Oh yeah. We have information about project success. We claim we hit schedule every time. But it's a lie. So the statistics looks good—but what we never admit is how many times we've moved the schedule. That's the real story—but we don't talk honestly about when we won't hit schedule and why. I can't think of a single project in the past two and a half years that was finished on the date initially planned." – *VP Information Services, manufacturing firm*

"This was not an IT person, and she was leading a project that required a lot of IT. The people below her weren't giving her all the information that they should have, and she wasn't able to connect the dots. They knew they were in trouble, but the project was at a high visibility. They just kept working hard—like they'd eventually get there.

"I took the project from green to red overnight, two weeks before it was supposed to go live. I had to involve the second- and third-highest persons in our company, and obviously that didn't go over well. The person who was the project lead—I felt bad for her."
– *Vice President, food services firm*

Impact of Speaking Up

The 26% who were able to address this problem** improved quality and functionality by 56% while reducing:

- Cost overruns by 55%
- Schedule delays by 69%
- Persistent problems* by 70%
- Inappropriate shortcuts by 66%

* Continued for the life of the project and beyond.
** Compared to those who didn't address the issue (these differences are all statistically significant, $p < .05$)

Application
THE ROUNDUP CASE

The enormous pressure to give good news set Kelley up for failure by making it difficult to discuss risks. It made it hard to admit that a single problem vendor could pose a risk to the entire project, even though the problems with the vendor were long-standing and numerous. It made it hard to admit that shortcuts during testing could pose a risk to the entire project, even though testing had been seriously limited to meet the aggressive release schedule.

The problem was that none of these risks was addressed either in Kelley's final project review or in prior sessions. Project participants had seen these problems coming for months, but no one had spoken up in an effective way.

PROBLEM 5

Too often, team members will ignore team deficiencies and work around the problem.

Extent and Costs

Of the 80% of study participants who reported the prevalence of team failures in their projects:

- 30% of their total projects were affected
- 76% said the issue was difficult or impossible to confront or resolve
- 49% spoke up at all
- 14% spoke up skillfully and had some effect on the problem
- 24% were able to solve the problem

80% reported that the problem persisted,* causing operational impact as follows:

- 73% of projects exceeded their budget
- 82% missed deadlines
- 77% had missing or wrong functionality or quality problems
- Team morale was damaged on 69% of projects
- 24% had a long list of problems to be resolved after the project ended

* Continued for the life of the project and beyond.

CRUCIAL PROBLEM #5:
TEAM FAILURES

Team members are unwilling or unable to support the project.

Project leaders generally have a common attribute: they have far more responsibility than they have authority. They "lead" large cross-functional groups with very few direct reports. Often, all have other masters—and it shows.

Project leaders report being hobbled by team members who don't show up to meetings, fail to meet schedules, or lack the competence to meet ambitious goals. Often these leaders have no say in selecting or replacing these nonperformers and feel powerless to coach them. Instead, they ignore their deficiencies and work around the problem.

For example, someone assigned to the team may be viewed by others as incompetent. Rather than addressing such concerns openly, additional resources are brought in to take over the most critical tasks. Not only is this a costly duplication of effort, it creates an environment of resentment and mistrust between the now-competing individuals and among other team members as they are dragged into this jockeying.

A second source of team struggle comes when project leaders must negotiate with functional managers to staff their projects. These functional managers often have many other priorities and are unresponsive to the project leader's needs. Once again, if project leaders can't hold effective crucial conversations with functional managers who fail to deliver, then project success is compromised.

Key findings include:

- When project leaders fail to effectively address performance problems in their teams, four out of five projects suffer from budget, schedule, and/or quality problems.
- Among the 80 percent of leaders who reported team member failures, observations about common causes include:

Where Team Members Let Each Other Down

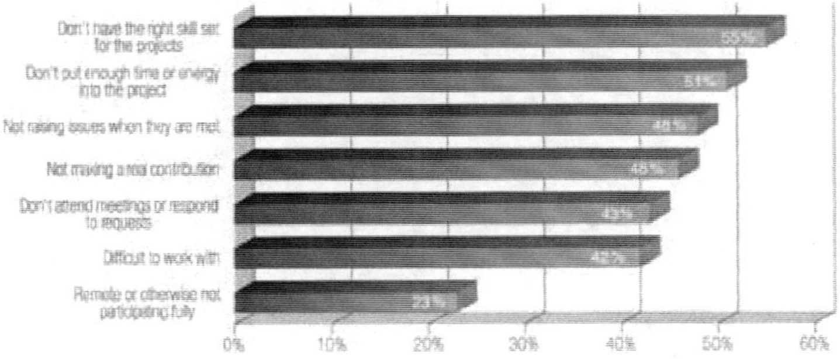

- 76 percent of those describe the discussion as between "difficult" and "impossible" to resolve.
- As a result, only 14 percent are able to skillfully raise their concerns in a way that gets heard.

With so few people able to succeed at this crucial conversation, it's no surprise that team failures begin in the earliest stages of projects and persist for their entire duration.

- 80 percent these projects suffer from team failures for the life of the project or beyond.
- Up to 82 percent of these projects go over budget, miss deadlines, and suffer from quality problems.
- More than two-thirds leave teams with battered morale.

Participant Comments

"The functional managers? You've got to hound them. I just had this one project. I kept reaching out to the manager of the resource group I needed. First I tried to invite them to the kickoff meeting so they'd see the scope of their involvement. Then I sent them private e-mails. I was getting blown off for weeks. I stop by their office, and if I ever brought up how they've been blowing me off, they really didn't like it. And escalate? The most my manager would do was send them an e-mail. No one really confronted the problem. That really stinks." – *Project Manager, food services firm*

"If your project isn't hot you won't get the resources they committed to you. They don't take you seriously until you have a crisis. When it becomes a crisis then we will address it. Now someone can step up and become a hero. Say the regulators are coming next week! Now you get the resources, the money, and away you go." – *Project Manager, financial services firm*

"Once they assign a bad resource you can only escalate to their manager. You just accept what they give you. Or just give their work to someone else. If that person has a track record, then obviously the manager knows, so why bother escalating? We do not have a way of turning down a resource or giving feedback to the manager or the individual. The functional manager has no idea how this person really functions. So you succeed by finding a horde of people you like and trying to synchronize the ending of one project with the beginning of your next so you can hang onto them." – *Project Leader, manufacturing firm*

Impact of Speaking Up

The 24% who were able to address this problem** improved quality and functionality by 64% while reducing:

- Cost overruns by 64%
- Schedule delays by 60%
- Persistent problems* by 70%
- Inappropriate shortcuts by 66%

* Continued for the life of the project and beyond.
** Compared to those who didn't address the issue
(these differences are all statistically significant, $p < .05$)

Application
THE ROUNDUP CASE

From the beginning, Kelley suffered from the common lot of all project leaders—he had far more responsibility than he had authority. He "led" a group of 1,000+ professionals—only seven of whom reported to him. All had other masters—and it showed.

Kelley, for example, was assigned a senior systems architect whom Kelley thought was incompetent. Rather than addressing his concerns openly, he brought in a second architect and made sure this person received the most critical tasks. Not only was this a costly duplication of effort, it created an environment of resentment and mistrust between the two architects and among other team members they dragged into their jockeying.

CONCLUSION

Application

THE ROUNDUP CASE

The complete undoing of Roundup was not guaranteed by the existence of these problems but by how they were handled. Most were never fully discussed with the people who needed to hear them. Others were raised, but then muffled by the very executives who should have listened and taken action.

KEY FINDINGS

The crucial issues being addressed are so common that many leaders have stopped seeing them.

Although *Silence Fails* focuses on important findings that can predict and explain failure, the most important implication of the study is that potential leaders have to influence success.

While the data shows that fewer than one in five project leaders effectively engage in the crucial conversations needed to solve these five problems, the good news is that one in five does. Much can be learned by studying their successes and whether these unique individuals produce different results than their stymied peers.

Research on this contingent of crucial conversations successes demonstrates that project leaders who speak up effectively do affect project success, and that others can be taught to speak up more skillfully and achieve similar results. Not only is it possible to rapidly and measurably change patterns of silence in an organization to patterns of open dialogue and accountability, it's also possible to sustain these results.[1]

Following are a few of the best practices project leaders and leaders can begin using immediately to address the issues raised in *Silence Fails*. Use these best practices to face the issues and begin building organizational competence at resolving them.

Getting People to Speak Up Well

As stated earlier in this report, in each of the five key problem areas, the study found a clear difference between speaking up and speaking up well. Across the problem areas, about half of leaders make some attempt to speak up. But most are ineffective. Some speak up but they water down their concerns, so the issues are never fully aired. Some speak up but do so in a way that provokes defensiveness from others. A handful are able to share their full concerns and by the end of the conversation feel their views are understood and respected.

The study also shows that while the skill of the initiator is a key ingredient in ensuring these crucial conversations are held well, the receptiveness of the other party is similarly important. Project leaders who are adept at holding crucial conversations are far more effective in dealing with prickly and defensive executives. But their skill doesn't guarantee that their concerns will be addressed. Unless and until leaders take extraordinary measures to ensure their environment is conducive to holding crucial conversations, a significant number of issues will remain unaddressed, invisible, and fatal.

1. Develop a Business Case for Change

The crucial issues being addressed are so common that many leaders have stopped seeing them. Rapid change can be made, but not if these crucial conversations are seen as "soft" issues—or as a "nice to do." Consider the following steps to develop your business case for change:

- Begin by making the problem visible. Track and publish data about project successes and failures.
- Distribute *Silence Fails* to generate discussion about the root cause of current underperformance.
- Engage senior leaders in a "listening campaign" where they lead structured focus groups to validate whether these crucial problems affect current results.

The goal of this process of published results, engaging discussion, and focused listening is to help senior leaders connect the behaviors described here with results they care intensely about. Once they can articulate this connection, they will see the business need for changing these vital behaviors.

2. Measure Behaviors

Leaders who are serious about improving how their people address these conversations are the first to regularly measure current behavior.

- Conduct a *Silence Fails* Assessment (find one at www.silencefails.com) to determine how well your project is performing in relation to the five crucial conversations in *Silence Fails.*
- Discuss the results of the *Silence Fails* Assessment with the right people. Whether it's the senior management team or another team of cross-functional leaders, this will help draw attention to the existence of the crucial issues and whether they are being adequately discussed and addressed.
- Use an organizational survey to assess executive competence within your organization.

KEY FINDINGS

When a skillful leader creates even a moderately safe environment, the likelihood of project success increases 50% to 70%.

3. Invest in Skills

Most project leaders and team members lack the confidence to address these politically sensitive issues because they don't know how to lead this risky discussion. Leaders who train their people to deal with these specific crucial conversations see substantial improvement in whether and how the issues get resolved.

- Publish a bold improvement goal. Get senior leadership committed to specific improvements over baseline measures.
- Follow up with training, discussion, assessments, and other tools to ensure that transference is occurring. Monitor the impact of this training in your organization.
- Map training outcomes to key performance indicators. Determine the behaviors that are connected to the results you want to improve and track both the behavioral change and that bottom-line result you're after.

4. Hold Senior Management Accountable

Investing in project leaders' competence at holding crucial conversations is necessary but insufficient. Effective organizations judge leaders and hold them accountable for improving the measures on this survey within their areas of responsibility. The best even tie it to senior leadership bonuses and performance evaluations. In addition:

- Make leaders the teachers. People will change their behavior more rapidly if leaders deliver the training than if staff trainers or outside consultants do so. When leaders teach, the speed of change can be two or three times greater than when those who aren't as credible and connected in the organization lead instruction.*

- Hold everyone accountable for creating a safe environment for these crucial conversations.

When senior leaders know that they are accountable for how well they lead at changing people's behavior, they get much more interested.

5. Make Heroes of Early Adopters

The key to getting a hundred people to speak up is to publicly reward the first one who does. Be sure to send a clear and public message that these conversations aren't just important, they're *crucial*—and those who raise them are highly valued. Tips include:

- Highlight and reward people who take a risk and raise these crucial conversations.

- In a public forum, praise individuals who surface sensitive project risks or skillfully challenge leaders.

If you want to change the values of an organization away from silence and toward candid dialogue, make heroes of those who take a chance with these new behaviors.

For the resources recommended here and additional tools, please visit www.silencefails.com.

Section Notes

1. Based on over twenty five years of research, VitalSmarts continues to conduct field studies that demonstrate significant, sustainable results brought about through organizational change initiatives. These efforts focus on behavior modification as opposed to system design and strategy development. See www.vitalsmarts.com for case studies and additional information.

VitalSmarts™

ABOUT THE STUDY

Beginning in late 2005, VitalSmarts and The Concours Group partnered to dive below the formal systems and identify what was missing. With worldwide participants, the study *Silence Fails* yielded substantial findings. The study was broken into two phases: exploratory and observational.

EXPLORATORY PHASE

Researchers conducted interviews and gathered statements from individuals in leadership positions, as well as individual contributors involved in project management. From there, the research team analyzed emerging themes and patterns to produce a set of issues that were most prevalent and costly. Results were retested with focus groups. Finally, a preliminary test of the survey was conducted to confirm that our factor analysis was consistent with a survey-based approach.

OBSERVATIONAL PHASE

The observational phase of our study, including a full survey, included the following:

- Researchers spent more than 150 hours observing various phases of project chartering, approval, and execution. In addition, interviews, focus groups, and surveys were administered to more than 1,000 senior executives, project sponsors, project leaders, and project participants.
- More than 1,000 executives and project management professionals across 40 companies were involved. Participating organizations came from a wide variety of industries, including pharmaceuticals, a fast food conglomerate, an international construction company, airlines, financial services, government agencies, and consumer products. While most of the 40 companies were Fortune 500 multinational organizations, about 10 percent were smaller, regional firms.
- The analysis encompassed more than 2,200 projects, ranging from $10,000 IT projects to billion-dollar organizational restructuring efforts.
- Study participants ranged in experience from veteran university-trained and certified project leaders to newly minted managers with little formal training in project management. Some organizations had sophisticated project governance, management processes, and policies, while others had far less developed approaches.

About the Sponsors

VitalSmarts

An innovator in corporate training and organizational performance, VitalSmarts helps teams and organizations achieve the results they care about most. With award-winning training products based on more than 25 years of ongoing research, VitalSmarts has helped more than 300 of the Fortune 500 realize significant results using a proven method for driving rapid, sustainable, and measurable change in behaviors. VitalSmarts has been ranked twice by *Inc.* magazine as one of the fastest-growing companies in America and has trained more than 500,000 people worldwide. www.vitalsmarts.com

The Concours Group

The Concours Group is a new breed of professional services firm, supporting senior executives through the blend of leading-edge intellectual capital and pragmatic business applications. The firm works with more than 300 of the Global 1,000 firms, helping leaders turn human and technological potential into business value. The Concours Institute is the research and education arm of The Concours Group. Concours research discovers and develops future best practices in business, technology, and human capital; its education articulates them and motivates their adoption; and its innovative senior-person Advisory Services enables clients to implement them quickly and achieve business results. www.concoursgroup.com

VitalSmarts™

APPENDIX

Aggregated study findings are outlined below.

		Fact-Free Planning	AWOL Sponsors	Skirting	Project Chicken	Team Failures
Extent of Problem	Project managers with this problem	85%	65%	83%	55%	80%
	Projects affected	39%	33%	33%	31%	30%
	Those reporting it's hard to confront/resolve	91%	88%	79%	61%	76%
	Those who speak up at all	61%	47%	51%	42%	49%
	Those who speak up skillfully	14%	11%	13%	13%	14%
	Those able to speak up in a way that solves the problem	19%	17%	16%	26%	24%
Cost of Not Speaking Up	The problem continues for the life of the project or beyond	88%	75%	81%	76%	80%
	Budgets are exceeded	82%	75%	78%	78%	73%
	Deadlines are missed	82%	85%	87%	86%	82%
	Functionality specs or quality suffers	82%	74%	80%	74%	77%
	Team morale is damaged	73%	69%	66%	54%	69%
	A long list of problems remain to be resolved after the project ended	43%	31%	28%	23%	24%
Impact of Speaking Up	Reduced cost overruns	29%	33%	35%	55%	64%
	Reduced schedule delays	17%	33%	28%	69%	60%
	Improved quality and functionality	69%	50%	44%	56%	64%
	Were less likely to take inappropriate shortcuts	47%	66%	64%	66%	66%
	Were less likely to have a long list of problems to resolve after the project ended	58%	70%	71%	70%	70%

For more information about this study, contact VitalSmarts at 800.449.5989 or via e-mail at research@vitalsmarts.com

VitalSmarts™

SILENCE FAILS ASSESSMENT
Will Your Project Succeed?

Consider a current project or a project that you've recently completed.

Project Name: ______________________________

Section 1: Fact-Free Planning

	Yes	No
1. Are the timelines, budgets, materials, and personnel established for this initiative realistic—given the deliverables required? (If you answer "Yes" on this question, go to the next section. If you answer "No," answer questions 1a–1c below.)	☐	☐
1a. If the resources are not realistic, have you fully discussed all of your concerns with the person(s) who could change them to make them realistic (e.g., project sponsors, resource managers)?	☐	☐
1b. Are you confident that the person(s) who can make the changes will take the actions required to adjust the resources, deliverables, or other elements of scope so that the initiative can succeed?	☐	☐
1c. If the person(s) who can make the changes to the resources or the deliverable or other elements of scope do so, will it be enough to make the project a success?	☐	☐

Score for Fact-Free Planning (add up the checks under "No"): ________

Section 2: AWOL Sponsors

	Yes	No
2. Is your sponsor providing the kind of active involvement and support this initiative needs to be successful (e.g., convincing reluctant stakeholders, supporting you in policy meetings, working behind the scenes to keep the initiative on track)? (If you answer "Yes" on this question, go to the next section. If you answer "No," answer questions 2a–2c below.)	☐	☐
2a. If you are not getting the level of support you need from your sponsor, have you fully discussed all of your concerns with him/her?	☐	☐
2b. Are you confident that your sponsor will soon become as actively involved as you need him/her to be?	☐	☐
2c. Is your sponsor's active involvement going to be enough to make the project a success?	☐	☐

Score for AWOL Sponsors (add up the checks under "No"): ________

Section 3: Skirting

	Yes	No
3. Are the right stakeholders supporting the priority-setting process with this initiative (e.g., not exerting inappropriate influence; not bypassing channels to change the scope, schedule, or resources; not pulling rank to get their way or cutting deals on the side in a way that threatens the initiative)? (If you answer "Yes" on this question, go to the next section. If you answer "No," answer questions 3a–3c below.)	☐	☐
3a. If a powerful person is exerting inappropriate influence on your project, have you discussed your concerns with him/her?	☐	☐
3b. Are you confident that this person will stop exerting his/her influence in inappropriate ways—e.g., bypassing channels, pulling rank, cutting deals?	☐	☐
3c. If this person stops exerting his/her influence in inappropriate ways, will that be enough to make the project a success?	☐	☐

Score for Skirting (add up the checks under "No"): ________

www.SilenceFails.com

Section 4: Project Chicken

	Yes	No
4. Are people giving you honest, accurate, and timely information regarding the status of your initiative, allowing you to get a clear view of where it stands? (If you answer "Yes" on this question, go to the next section. If you answer "No," answer questions 4a–4c below.)	☐	☐
4a. If a person is failing to get you honest, accurate, or timely information on the status of the project, have you discussed your concerns with him/her?	☐	☐
4b. Are you confident that this person will start getting you honest, accurate, and timely information on the status of the project?	☐	☐
4c. If this person starts getting you honest, accurate, and timely information, will that be enough to make the project a success?	☐	☐

Score for Project Chicken (add up the checks under "No"): ________

Section 5: Team Failures

	Yes	No
5. Are team members performing up to your expectations and keeping commitments and are they sufficiently competent to see the initiative through to the end? (If you answer "Yes" on this question, follow the instructions below to score your assessment. If you answer "No," answer questions 5a–5c below.)	☐	☐
5a. If a person is not performing up to your expectations (i.e., he/she does not have the right competencies, talents, time, etc.) or if a resource manager is not assigning qualified candidates, have you discussed your concerns with him/her?	☐	☐
5b. Are you confident that this person will begin performing up to your expectations?	☐	☐
5c. If this person begins to perform up to your expectations, will that be enough to make the project a success?	☐	☐

Score for Team Failures (add up the checks under "No"): ________

Section Score:

The section with the highest number of "No" responses is your weakest section. Overall Score (add up the total score from each section):

0: Your initiative has a good chance of success.

1–4: Your initiative has a few issues, but if you deal with the problem areas, there is still a good chance that your initiative will succeed. Unfortunately, If you fail to address even one of these crucial conversations, there is a 61 percent chance that your project will go over in budget, schedule, and scope.

5–9: Your initiative is in jeopardy. You have one or more big problems and possibly several smaller problems that you need to address to get your project back on track.

10–14: You have three to four major conversations that you need to hold and hold well.

15–20: Speak up! Your initiative is doomed!

To find out how to get better, or to request a team or organization assessment, visit www.silencefails.com

Next Steps for Leaders

Start predicting and preventing the failure of your high-stakes business initiatives

For more than twenty-five years, the founders of VitalSmarts have researched methods for bringing about systematic and lasting change. Based on this ongoing research, VitalSmarts has helped thousands of organizations, including more than three hundred of the Fortune 500, realize quick, hard-hitting results.

From this extensive research, we've gleaned high-leverage tips to help you transform your organization:

- Download a PDF version of the *Silence Fails Assessment* at www.silencefails.com. This assessment highlights the points at which your initiative, program, or project is likely to fail.
- Distribute this report to help others become aware of the problems it reveals.
- Present the *Silence Fails Executive Summary Presentation* featuring acclaimed speaker Joseph Grenny to your team. Joseph will walk you through the insights revealed in the study and show you how you can achieve flawless execution. (Available in this kit and at www.silencefails.com.)
- Invite Joseph or one of many other noteworthy speakers from VitalSmarts into your organization to generate awareness of the five crucial conversations for flawless execution.
- Once you've increased awareness and appreciation for the issues, it's time to move to action. A VitalSmarts Representative can help you develop an action plan that will meet your specific needs. Call 1.800.449.5989 today to find out how you can begin to change your organization.

Other tips and tools for change are available in the conclusion section of the research report and at www.silencefails.com.

Silence Fails

Catching Project Problems Early

By Stephen L. Cabano
President – Pathfinder, LLC

Abstract

This paper will address these issues as well as provide guidance as to how this relationship can be initiated and nurtured throughout the life of the project.

This paper will address many of the tools and techniques that are available today to help project managers predict the future. Project planning techniques such as Project Execution Planning (PEP), Cost and Schedule Risk Assessments, Independent Project Reviews (Cold Eye Reviews), etc. will be addressed. Project Control features, such as progress measurement, Earned Value, productivity, trending and forecasting, will also be addressed. Although the industry has proven these tools to be value-added, they are useless to the success of a project unless the project manager acts on the information these tools provide.

This paper will address projects in both the US and International Chemical Process Industry. Case study examples will be used to demonstrate the benefits of these proven project management tools and techniques. Lessons Learned from a number of seasoned project managers will be conveyed to reinforce the benefits of following sound project management best practice.

Introduction

Wouldn't it be great if you had a crystal ball that could predict project problems before they happen? Experienced project managers have such a capability, and in most cases these skills have been developed by years of executing projects and observing/experiencing the same project problems repeatedly. The industry has spent years developing project management "Best Practices" that provide the basis for identifying project issues early and providing the project team time to react to the problem.

Most Chemical Engineers are trained to engineer a solution that will allow the end users to achieve their business needs. These typically result in choosing the best technology, developing the project drawings and support specifications, and completing a design package that will meet the end users' needs. This provides the "What" that needs to be built. Project Management is focused on the "How." How do we design and build it? Many Project Engineers are not trained or experienced in the "How," which results in inefficient and ineffective execution of the project. The "How" includes planning, estimating, scheduling and control of the actual work. These are the features that allow us the ability to look into the future and minimize issues that may negatively impact project outcome. If the "What" and "How" are not aligned, the project can experience numerous changes which result in longer execution time and costlier facilities.

Projects in today's world will not be successful unless there is close communication between the owner (who will operate the facility when completed) and the contractor (who will design and build the facility). Many project issues arise when this relationship breaks down – the contractors are not specifically told what is expected of them, the owner does not manage the contractor, the contractor performs poorly, or the owner constantly changes his mind on the project scope and/or schedule.

Industry Perspective

As we all know, the world is experiencing an economic crisis unlike any we have ever witnessed before. This has affected the capital project planning and execution environment dramatically. Pricing, contractor availability, vendor activity, material cost, etc. are all affected by the world economy. Owner organizations are also reconsidering their overall capital spending which has resulted in numerous project cancellations and slowdowns.

Over the past 5 to10 years, the capital project planning and execution industry has experienced a huge human resource availability issue. Many of the existing project professionals are reaching retirement age and opting to take early retirement, which has left the industry with a shortage of experienced project management talent. This is a great opportunity for young project representatives to move up the project management career ladder faster than their elder counterparts were able to do. This can be a challenge, though, when one moves into a leadership role with limited experience and lacking the Lessons Learned background.

These issues have impacted owner organizations as well as contractors, vendors and suppliers. This leads to a lack of experience and knowledge across the industry and can result in poor project guidance and oversight with limited outlets for advice and consultation. So, how does a young engineer excel in this industry, recognizing the issues outlined above and the challenges looking forward?

Attributes of Highly Effective Project Managers

One of the greatest attributes a project manager can have is "knowing what you don't know", and in response, surrounding himself/herself with people that do have the knowledge and expertise and "do know". These people can be sourced from internal project resources, contractors, vendors and/or consultants. These experienced team members will provide support by identifying project risk early and developing proven mitigation strategies that will eliminate the risk or, at least, minimize the impact to your project. Another key attribute is the ability to make the team function as efficiently and effectively as possible. Trying to optimize everyone's work environment can be done easily, but a project manager needs to recognize that, based on the defined business and project drivers, some inefficiencies cannot be avoided. For example, if a project is schedule driven (the project needs to be completed faster than the competition in order to capture the market need), we may need to keep additional engineering resources on staff longer to address any late changes in design. This may come at a financial cost to the project but, if speed is of the utmost importance, this inefficiency may be the best strategy for the business and the project.

Other key attributes of an effective project manager are that he/she needs to be extremely organized, able to juggle multiple tasks at any time, resilient when addressing issues that do not go as planned, must be able to always see the big picture when addressing problems, must rely on the support of the project team and their ability to execute the tasks required, and must be able to communicate to upper management, as well as across his/her project team. To make all of the above possible, the Project Manager needs to be a planner – one who looks into the future and sees the end goal and defines the steps required to get there. (See Figure 1)

Typical Project Envelope

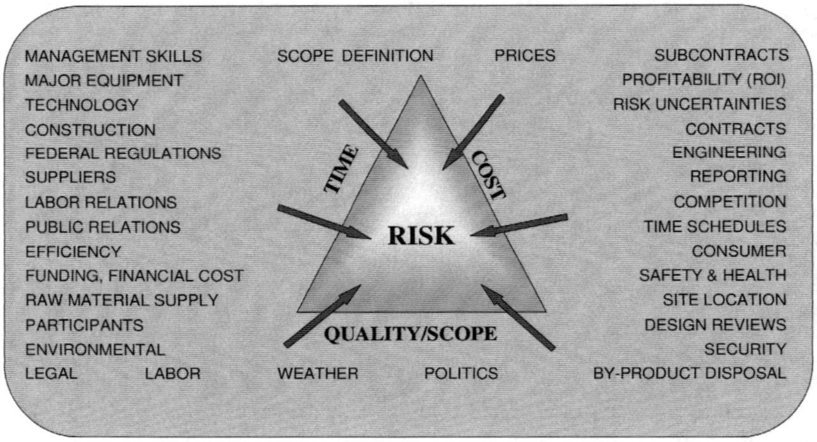

Fig. 1

One of the biggest issues facing quality project managers is that they need to relinquish their desire to "do the work". Most Project Managers in the Chemical and Hydrocarbon process industry came out of the technical fields (Chemical Engineer, Mechanical Engineer, Construction representative, etc.) and excelled at these positions. They still want to do the work that they were comfortable with, which draws them away from the managerial responsibilities of all the project pieces. A project manager needs to oversee the people doing the work, recognize when things are not going as planned, be willing to take action to correct any problems and capture any improvement opportunities.

The above are just a few of the characteristics of leading project managers in the industry. Not everyone is cut out for Project Management responsibilities, as some people like to "do the work" rather that oversee. We often try to force fit our most senior and experienced engineers into managerial roles which may result in a double disaster: 1) the individual does not function well in the Project Manager role and 2) you've lost one of your best engineering resources. Management needs to recognize this and have a well-defined and well-implemented project manager selection process.

The Planning Process

The project risk identification and risk mitigation process starts very early in the project planning process. As soon as the business opportunity is identified and management agrees to invest in the front end engineering and planning to capture the opportunity, a project manager should be assigned. The project manager's role in these early phases is to evaluate technical and execution alternatives and select the best alternative to meet the pre-defined business and project objectives. Every engineering solution and related execution strategy has its benefits and challenges. The project manger's role is to weigh these plusses and deltas against the business needs and ultimate project desires, so the team will be able to select the optimum project solution. As identified above, this is when the project manager needs help. He/she needs to draw from knowledgeable people and expert resources in order to lay out all the options and, in turn, identify what is optimum for the specific project.

The Project Management Process

BUSINESS PLANNING (1)	ALTERATIVES ANALYSIS (2)	BASIC ENGINEERING (3)	EXECUTION -EPC- (4)	OPERATE & EVALUATE (5)
OBJECTIVES:	OBJECTIVES:	OBJECTIVES:	OBJECTIVES:	OBJECTIVES:
• Clearly frame goal.	• Generate alternatives.	• Fully define scope.	• Implement Execution Plan.	• Monitor performance.
• Identify opportunities.	• Reduce uncertainty and quantify associated risks.	• Develop detailed execution plans.	• Finalize Operating Plan	• Benchmark performance against objectives and competitors.
• Test for Strategic fit with business objectives.	• Develop expected value for selected alternatives.	• Refine estimates & economic analysis to A/R level.	• Collect, analyze, and share metrics & lessons learned.	• Share results and lessons learned.
• Preliminary assessment of uncertainties, potential return and associated risks.	• Identify preferred alternative(s).	• Confirm if expected value meets business objectives.		• Continue performance assessment and identify opportunities.
• Plan for next phase.	• Plan for next phase.			

Fig. 2

Once this selection is made, the Project Manager's role is to define the selected technical and execution options in enough detail to allow management to make a decision whether or not to invest in the execution of the project. This includes the identification of a clearly defined scope of work, its supporting cost estimate and schedule, a well-document execution plan and a clear understanding of the risk associated with the project moving forward. This will allow the manager to make an educated decision on investing further in the project and the risks associated with the investment.

Relying on Others

As stated numerous times now, the project manager cannot execute a project alone. As our project workforce ages, younger, less experienced project engineers are being asked to fill the role of project managers. These resources are highly talented, energetic, hard-working and even bring some innovative ways of communicating and reporting, but lack the "been there, done that" experience. These individuals need to rely on their communication skills to dialog with the remaining experienced workers who can provide some of the Lessons Learned. Many organizations are pooling these highly experienced resources into a 'Center of Excellence Group' which provides project oversight and assurance that proper project considerations are followed.

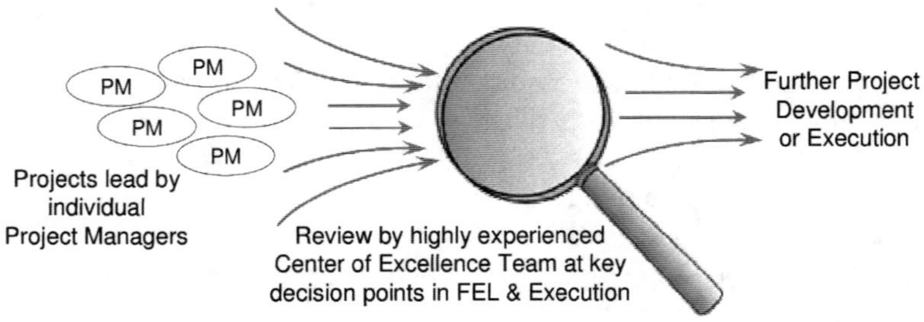

Fig. 3

Organizations need to consider these internal talent pools as valuable project contributors rather than project auditors. The negative connotation of an audit team deters organizations from bringing them in, even though they could provide honest, value-added contributions. The innovative and creative minds of the younger project representatives combined with the experience and knowledge of the more 'seasoned' team members should result in more efficient and effective project solutions for today's world.

Project Execution Planning

One of the most critical aspects in any project is a well thought out and documented project execution plan. This plan addresses the integration of all the project aspects such as scheduling, contracting, team roles and responsibilities, funding, site logistics, procurement requirements, communications, reporting, etc. The plan is comprehensive, concisely documented and communicated to all team members. It assures that all project team members are clear regarding their responsibilities and interfaces.

This plan also defines the baseline expectation for the team. All project control will be based on these baseline parameters. The team will need to address any deviations from the plan whether positive or negative.

Fig. 4 Project Execution Plan

Deviations from the plan are early indicators of project issues, since the plan was developed by incorporating all of the input from the project team members. It provides the "Crystal Ball" that we are looking for in support of project execution. Project control progress measurements and productivity will be established by the baselines set in this plan.

The Tools

There are a multitude of tools that support the overall project management and project control process. The following address the ones that can help us predict the future, or at least give a glimpse of what might occur based on history and trends:

- **Project Benchmarking** – this technique is widely used across the process industry. It allows us to evaluate how we are doing against others in the industry with similar projects. There are organizations that provide these services as well as self-assessment tools that are then measured against a database of similar projects.
- **Industry Rating Indices** – these indices are used in support of the above outlined benchmarking efforts. The indices allow project teams to measure how much detail has been defined prior to project authorization and, based on history, your probability of meeting your defined cost and schedule objectives.
- **Lessons Learned Databases** – organizations have realized that we encounter similar issues repeatedly across projects. The Lessons Learned databases allow us to review past project issues and plan around these issues so our end goals can be achieved.

- **Appropriate Contracting Steps** – in order to gain quality support from our contractors, we must be sure that we complete fair and equitable contracts that have well-defined job specifications and sound, efficient and comprehensive Project Instructions. This will allow the contractor to bid effectively and produce quality deliverables.

- **Clearly Defined Roles & Responsibilities** – the project manager needs to assign specific roles and responsibilities to all team members so nothing is overlooked and tasks are not overlapping across team members.

- **Risk Analysis Systems** – the industry has developed excellent risk analysis systems that allow the project team to evaluate potential execution vulnerabilities to the project and their cost and schedule impacts. If the impacts are too great, the team decides on mitigation strategies to minimize their impact.

- **Independent Project Review Support** – an excellent resource for project support is an independent review of the project by highly experienced project professionals. Some organizations have this capability internally and some use outside resources. If the review team observes possible project issues, the project team is made aware of these issues and decides on a course of action to mitigate any possible negative outcomes.

- **Peer Assistance** – many organizations have a mechanism (internal to their project process) that allows the project team to call in experts at key points in the process to assist in the development of project related deliverables. These might include expert construction planners, cost estimating assistance, start-up experts, etc.

- **Progress/Productivity Measurement** – the industry has developed a series of project control tools and techniques which allow the project team to take information collected on the project and project the future, based on this experience. These trends allow us to catch project deviations early and adjust the execution approach accordingly, to mitigate the issues.

The above are just a few of the tools and techniques that support our ability to look into the future and provide us with the information to make decisions regarding team composition, supervision, training, execution strategies, etc. Capturing these issues early offers us a number of optional resolution solutions and mid-course corrections if needed. If we wait too long to address these issues, our options are minimized and project success is jeopardized.

International Projects

When executing international projects, the need for experienced support is heightened. The team needs information on the various aspects that will impact the project. These issues include:

- **Cultural differences** – work hours, communication lines, and working environments may all differ from one location to another.
- **Time zones** – the project team may be dispersed across multiple time zones, which needs to be considered when communicating and planning meetings.
- **Currency exchange rates** – for international projects and in most domestic projects today, materials, equipment and event labor may be purchased across country boundaries and paid for in different currencies.
- **Government rules and regulations** – the rules and regulations established by local governments may differ dramatically.
- **Local labor capabilities** – some local governments may mandate use of local labor and labor capabilities may also differ from region to region.
- **Infrastructure issues** – when executing projects in different countries, the local road systems, phone capability, electrical reliability, etc. may not be as stable as in other countries.
- **Socio-Political Issues** – the local governments and populations may have political, religious and other differences which may cause unrest in the region.
- **Safety and Environmental Guidelines** – these guidelines differ from state to state in the United States and also differ dramatically from country to country.
- **Approved lines of communication** – international companies and governments may have established communication requirements that differ from those of your company.
- **Payments protocols and nuances** – when dealing on an international scale, payment terms and invoicing requirements may differ dramatically from country to country.

None of these issues are game breakers but, if they are not addressed, can wreak havoc on your project. Effective utilization of the previously outlined tools and having experienced representatives who know how to deal with the above mentioned regional issues can help organizations capture any problems early and pay huge dividends on your project's success.

Conclusion

So, with the tools and techniques outlines above, we might not be able to predict the future but we can catch many project problems early. The earlier we foresee these issues, the better we can manage them and the more support we will receive from all project stakeholders. Will this assure that all projects are delivered on time and under budget? Probably not, but it will create an environment where you are in much better control of your projects, more informed decisions are made and the probability of increased project success is much greater. Employing these tools and techniques will result in better utilization of the company's capital money, higher levels of confidence on project outcomes and increased reliability on the project management discipline.

What Does It Cost?

Richard F. Garnett, Honda Manufacturing of Alabama, Lincoln, AL

The first question a manager asks about a new project is: "What does it cost?" Engineers are well schooled in answering questions about design and technical performance, but are often less well prepared to answer the "How Much?" question.

This lecture deals with the complexities of capital approval, which is a technical, economic, and political process. The lecture reviews existing cost estimating techniques and discusses the level of effort required to generate a detailed +/-10% cost estimate.

It includes a discussion of the common mistakes engineers make in dealing with managers on cost and the importance of successfully navigating the "What Does It Cost?" question.

Presented in an interesting, fast pace style, and illustrated with real life stories the talk is suitable for a general engineering audience.

See more of this Session: Project Managment -- How to Overcome the Problems
See more of this Group/Topical: Management Division

Managing A Successful Project - Owner Vs Contractor Approach

Syamal K. Poddar, Member F&PD and Management Division, Poddar & Associates, Houston, TX

For any given project, from the initiation or conceptualization to the completion of the project there are two groups involved. These two groups are:

1) The Owner most often an operating company and

2) The contractor who is selected by the owner to execute the project with a well defined scope of work (with defined budget and schedule)

- Owner's View:

  Most often gold plated

  Drive the cost down

- Contractor's view:

  Deliver a functional plant

  Maximize the profit

- The Challenge is to bring a unified view for an acceptable common goal

Depending on the scope of work and the type of contract in existence, the involvement of the owner with the contractor can vary from a minimum level to a substantial level of involvent to drive the project to a successful completion. Often the interest and driving forces of the above two groups may be in conflict

In this paper the criteria of a successful project will be examined together with highlighting of the potential conflicts between the two groups. This will then be followed by examining how an alignment of the attributes and driving forces of the two groups when achieved from the very outset could lead to a successful execution of any project.

Owner's perspective:

Owner (an individual company or a consortium of companies) wants to have his project completed successfully within allocated budget as well as within the stipulated time period at the same time keeping his risk exposures to a minimum. The scope of work is developed and established by the owner's team. Sometimes the owner appoints an outside consultant to help his team. Such a consultant often known as Project Management Consultant (PMC) gets integrated with the owner's team and guide the project in the initial phase of the life cycle (firm inception to project scope definition stage The involvement of PMC (often considered as a part of the owner) is there to protect the interest of his client, the project owner. In that scenario the PMC stays as a PMC to develop the project working hand in hand with owner's representative and usually they get involved from the inception to the stage of developing the project scope book to clearly defining the scope of work.

Occasionally depending on the existing relationship, the designated PMC is also awarded the contract to execute the project. Although there is apparent notion of favoring the PMC company. There are several advantages of such an approach: The advantage of doing that is to cut down the cycle time of preparing a bid, inviting the bid response and selecting the suitable contractor following a specified selection criteria. ng the suitable contractor following a specified selection criteria. Often this approach helps the contractor to offer a better price because the contractor knows the scope better than an outsider who understands from the bid package and through bid clarification meeting. Also it eliminates the cost of preparing the bid response document and associated real time to do so. Therefore, overall there is a significant real time savings to shorten the overall project execution schedule. Also because of the coherent nature of the inter relationship between the contractor and the owner, the risk sharing becomes acceptable by the PMC when it is selected by the Owner to execute the EPC project. This is am effective way to launch a lump sum turn key project.

Contractor's Perspective:

The overall objective of the contractor is to execute the project to a successful completion where it leaves behind a happy client for the future repeat business. The inherent intention of the contractor is to maximize its own profit and to minimize the associated risk.

The execution strategy and motivation to do a good job are depended on the type of contract that drives the project.

Type of contracts: Basically Four Types

- Lump – sum or Fixed price (Hard – money)
- Reimbursable
- Reimbursable with fixed fee
- Unit price

The characteristics of each of these types of contract are summarized in the table 1 below:

TABLE 1

Criteria	Lump sum	Reimbursable	Reimbursable with fixed fee	Unit price
Pricing	Highly competitive	Competitive	Competitive	Competitive
Contractor risk	High	Lower	Moderate	Low
SoW definition	Detailed definition, fixed scop	General definition, variable scope	General definition, variable scope	Semi-detailed definition variable scope
Client risk of cost overrun	Lower	Higher	Moderate	Moderate

It is apparent from Table 1 that from the risk point of view the reimbursable with fixed fee type contract is equitable for both owner as well as contractor.

A better alignment of project owner and contractor is achievable by introducing incentive scheme in such a contract.. There are two examples shown in Figure 1 where in example A, the contractor is incentivized by earned bonus for early completion where as in example B, there is positive incentive (bonus) for early completion which is balanced by negative incentive (penalty) for schedule delay.

As shown Figure 1, by challenging the contractor with an incentive plan (preferably with a scheme shown as example B above), the contractor and owner are made to focus more objectively

Examples of Incentive Plans in a Contract

Objective: Alignment of client
& contractor expectations

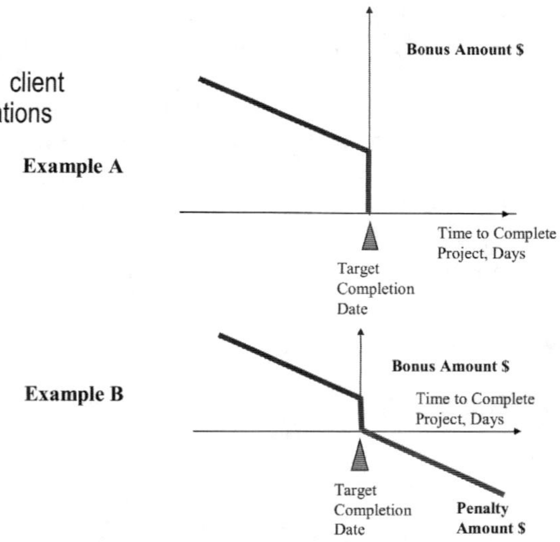

FIGURE 1

What is the best strategy to minimize (if not totally eliminate) the potential conflict between C (Contractor) and O (owner)?

The answer is simple to visualize yet challenging to institute. The first step is to recognize the typical barriers. Recognizing them then both parties need to work together to overcome them. The typical barriers are highlighted below:

Typical Barrier for alignment between C &O:

Planning is a waste of time and money We know how to do it, been there done that (btdt), no room for innovative ideas

Personal preferences vs. project goals

Lack of construction, operations, and maintenance (key stakeholders) participation

Contractor and/or owner culture

Lack of proper Owner/Client Input/Involvement/ Buy-in

Poorly defined/communicated requirements/ expectations (what owner wants and why)

Owner operations, maintenance or other representatives not "on-board"

Lack of proper attention to Project "Team building"

Contractor internal

Client/Contractor

Client internal

Lack of adequate planning

Start work before establishing plan

Start work before Project Team buys in and commits to the plan

Insufficient commercial focus

The deal (contract)

Financial considerations

Risk mitigation

Change control

Poorly defined/communicated scope basis

Work activities improperly planned/scheduled

Responsibilities not clearly defined/communicated

Project not properly organized/staffed

Work activities not properly managed/coordinated

Information flow not properly managed

Project Controls not properly established/utilized

Reactive rather than proactive management

Concluding remarks:

- A Project is successful only when the project owner is satisfied
- While the Owner drives to achieve "money's worth" the Contractor needs to make it profitable

---

Abstract ID#: 162641
Password: 693067
Title: Managing A Successful Project - Owner Vs Contractor Approach
Session Selection: 05000 Project Management -- How to Overcome the Problems
Preferred Presentation Format: Oral
Lead Presenter's Email: associates.poddar@gmail.com

First author

Presenting Author
Syamal K. Poddar
Principal Consultant / President
Member F&PD and Management Division
Poddar & Associates
14923 Wilderness Cliff Ct.
Houston, TX 77062
Phone Number: 713 494 3050
Email: associates.poddar@gmail.com

Overview of the Project Management Processes in the 4th Edition of PMI®'s PMBOK® Guide

(Throughout this paper PMI®, PMP®, and *PMBOK® Guide* are registered trademarks of the Project Management Institute)

Author: Dr. Eldon R. Larsen, PMP®
Marshall University
100 Angus E. Peyton Drive
South Charleston, WV 25303

INTRODUCTION

In 2009, PMI® (the Project Management Institute) published its fourth edition of *A Guide to the Project Management Body of Knowledge*, also known as the *PMBOK® Guide*. In the fourth edition, refinements were made that improve and better focus some of the processes needed for effective project management.

The *PMBOK® Guide* is a guiding summary to much of the knowledge needed for project management, which knowledge overlaps other management disciplines, but it does not cover all project management knowledge—a good project manager needs to know more than is contained in the *PMBOK® Guide*.

Before we launch into this brief overview of the project management processes contained in the *PMBOK® Guide*, let's first define what we mean by the terms *project* and *project management*. A *project* has specific, unique objectives and is time limited, meaning it is temporary. *Project management* is focused on identifying and addressing project and stakeholder requirements, while balancing and managing within the project constraints. The three major keys to successful project management are deep understanding and effective use of (1) the importance of people, (2) the importance of defining project objectives, and (3) the importance of good planning. The *PMBOK® Guide* does a very good job of using these three major keys.

The *PMBOK® Guide* is a summary of information that is generally recognized as being good practice on most projects most of the time. It provides a summary and standardization of common project management terms and processes to aid project managers in applying the knowledge, skills, tools, and principles of project management.

The *PMBOK® Guide* is a useful reference for people who are preparing for PMP® certification. It frequently serves as a resource for project management training, and it is used as a reference for accreditation of educational programs.

The *PMBOK® Guide* consists of the following project management knowledge: project life cycle definition, five project management *process groups*, nine different *knowledge areas*, and forty-two project management *processes*.

Project life cycle definition generally embraces the work to accomplish in each project phase, the deliverables to be completed in those phases, the criteria that each deliverable must meet, who is to be involved in each phase, the criteria for passing from one phase to the next, and who will make the decisions regarding those inter-phase passages.

Project management applies, integrates, aligns, and iterates as needed a total of forty-two project management processes within the following five process groups: (1) *Initiating*, (2) *Planning*, (3) *Executing*, (4) *Monitoring and Controlling*, and (5) *Closing*. Additionally, there may be product-oriented processes that must be further integrated with the project management processes.

The *PMBOK® Guide* defines a project management *process* to be a combination of interrelated actions focused on achieving specified results. These processes will likely need to be iterated, or even revised, multiple times to define and achieve stakeholder requirements and reach agreement on the desired process outcomes. The project management processes appear to be separate, but actually often overlap. Thus, they are interdependent with other project management processes, making it necessary that the project manager manage these process interactions to attain success.

It is important to note that the five project management *process groups* are *not* the same as project life-cycle phases. There can be many or few project life-cycle phases, and much of the work in the project management *process groups* is concurrent rather than in phase-like series. In fact, for a very large project, all five *process groups* might be used for *each phase of the project*.

The *processes* within the *process groups* each have what is referred to as *process elements*, called *inputs*, *tools and techniques*, and *outputs*. Below are listed some example, typical process inputs, tools and techniques, and outputs.

Some typical process *Inputs*

- Enterprise Environmental Factors
- Organizational Process Assets
- Project Management Plan
- Project Scope Baseline
- Project Scope Statement
- WBS
- WBS Dictionary
- Work Performance Information
- Performance Reports

Some typical process *Tools and Techniques*

- Expert Judgment
- Project Management Information System
- Configuration Management System
- Change Control System

Some typical process *Outputs*

- Change Requests
- Work Performance Measurements
- Updates

OVERVIEW OF THE FIVE PROCESS GROUPS

Initiating

In the initiating process group the following occurs: The project charter is developed; the project manager is named for the project; the project or project phase is authorized; and the project or phase is started.

The *initiating* process group consists of the following processes:

- Develop Project Charter
- Identify Stakeholders

Planning

In the planning process group the following occurs: The project scope is defined; the project objectives are defined and refined; and the project management plan is developed so that the actions needed to achieve the project objectives and project scope clearly identified.

The *planning* process group consists of the following processes:

- Develop Project Management Plan
- Collect Requirements
- Define Scope
- Create WBS
- Define Activities
- Sequence Activities
- Estimate Activity Resources
- Estimate Activity Durations
- Develop Schedule
- Estimate Costs
- Determine Budget
- Plan Quality
- Develop Human Resource Plan
- Plan Communications
- Plan Risk Management
- Identify Risks
- Perform Qualitative Risk Analysis
- Perform Quantitative Risk Analysis
- Plan Risk Responses
- Plan Procurements

Executing

In the executing process group the following occurs: The people and other resources are integrated to do the work of the project management plan, for either the entire project or for a project phase.

The *executing* process group consists of the following processes:

- Direct and Manage Project Execution
- Perform Quality Assurance
- Acquire Project Team
- Develop Project Team
- Manage Project Team
- Distribute Information
- Manage Stakeholder Expectations
- Conduct Procurements

Monitoring and Controlling

In the monitoring and controlling process group the following occurs: Project team members, and/or other people assigned to support the project, measure and monitor project progress in order to identify variances from the project management plan, so that corrective actions can be taken to meet project or phase objectives and performance criteria.

The *monitoring and controlling* process group consists of the following processes:

- Monitor and Control Project Work
- Perform Integrated Change Control
- Verify Scope
- Control Scope
- Control Schedule
- Control Costs
- Perform Quality Control
- Report Performance
- Monitor and Control Risks
- Administer Procurements

Closing

In the closing process group the following occurs: Formal acceptance of the product, service, or result is achieved, and the project *or a project phase* is brought to an orderly end.

The *closing* process group consists of the following processes:

- Close Project or Phase
- Close Procurements

Note that a project is not completed until all of the following are done:

- Customer acceptance—Product is verified against scope & delivered to customer's satisfaction
- Contract is closed out
- Project records are updated, project archives and lessons learned are updated, and team is released

OVERVIEW OF THE NINE KNOWLEDGE AREAS AND PROFESSIONAL RESPONSIBILITY

The *PMBOK® Guide* identifies nine basic *knowledge areas* that are essential to successful project management. In addition to the nine knowledge areas, the guide introduces the importance of taking *professional responsibility*. Here is a summary of each of the nine basic knowledge areas.

Integration Management

Integration management is primarily concerned with effectively integrating the forty-two project management processes, to be discussed later in this paper, to achieve the project objectives within the organization's procedures. It ensures that all the project processes work together, that the project charter and scope are developed, and it ensures that the project management plan is developed and appropriately implemented. Finally, this knowledge area helps the project to properly interact with the performing organization.

Integration Management consists of the following processes

- Develop Project Charter
- Develop Project Management Plan
- Direct and Manage Project Execution
- Monitor and Control Project Work
- Perform Integrated Change Control
- Close Project or Phase

Scope Management

The main purposes of scope management is to ensure that all required project work, and only the required work, is done to complete the project successfully. Scope management defines and controls what is in the project and what is not.

Scope Management consists of the following processes

- Collect Requirements
- Define Scope
- Create WBS
- Verify Scope
- Control Scope

Time Management

The purpose of time management is to develop and control the project schedule so as to complete the project within the required time schedule. Here, "Time Management" means "Schedule Management" and has nothing to do with personal time management, which is very important, too.

Project Time Management consists of the following processes

- Define Activities
- Sequence Activities
- Estimate Activity Resources
- Estimate Activity Durations
- Develop Schedule
- Control Schedule

Cost Management

The purpose of cost management is to estimate the project cost and to complete the project within the approved budget. It includes processes to (1) estimate monetary resources needed to complete the project, including schedule activities and outsourced work; (2) estimate a budget by combining the estimated schedule activity and/or work package costs to establish an authorized cost baseline; and (3) control costs by monitoring the project status, controlling cost variances, and controlling any cost baseline changes or updates.

Project Cost Management consists of the following processes

- Estimate Costs
- Determine Budget
- Control Costs

Quality Management

The purpose of quality management is to determine the quality policies, objectives, and responsibilities to achieve project and product requirements and objectives. Here we define *quality* to be the degree to which project and product requirements and objectives are satisfied.

Project Quality Management consists of the following processes

- Plan Quality
- Perform Quality Assurance
- Perform Quality Control

Human Resource Management

The purpose of human resource management is to obtain, organize, develop, lead, and manage the project team. The team members have assigned project roles and responsibilities. The number of and characteristics or types of team members can change throughout the project. It is generally important for team members to help with planning and decision making; the earlier the better.

Project Human Resource Management consists of the following processes

- Develop Human Resource Plan
- Acquire Project Team
- Develop Project Team
- Manage Project Team

Communications Management

The purpose of communications management is to generate, collect, distribute, store, retrieve, and archive project information in a timely and appropriate manner. Project managers spend the majority of their time communicating with project team members and other stakeholders!

Project Communications Management consists of the following processes

- Identify Stakeholders
- Plan Communications
- Distribute Information
- Manage Stakeholder Expectations
- Report Performance

Risk Management

The purpose of risk management is to identify positive and negative risks and how to respond to them if they occur and/or how to prevent negative risk impact and enhance positive risk impact.

Project Risk Management consists of the following processes

- Plan Risk Management
- Identify Risks
- Perform Qualitative Risk Analysis
- Perform Quantitative Risk Analysis
- Plan Risk Responses
- Monitor and Control Risks

Procurement Management

The purpose of procurement management is to manage the acquisition of products, services, or results from outside the project team in order to complete the project. Here we define a *vendor* as anyone who provides the product, service, or result is the seller; and we define the *seller* as the purchasers of the product, service, or result is the buyer. The project's organization could be a buyer and seller.

Procurement Management consists of the following processes

- Plan Procurements
- Conduct Procurements
- Administer Procurements
- Close Procurements

Professional Responsibility

Project management team has a professional responsibility to its stakeholders, including customers, the performing organization, and the public. Professional project managers and PMI members must agree to adhere to the PMI *Code of Ethics and Professional Conduct*.

Following is a summary of the *Code of Ethics and Professional Conduct*. Each PMP® has a professional responsibility to customers and the public. He or she

- Meets standards and requirements with integrity and honesty in all ways
- Conducts all business, estimates, work, communications, etc., with integrity and honesty
- Is accountable and responsible
- Keeps confidential information confidential

The PMP® is in essence working for the customer. The PMP® avoids conflicts of interest, either real or apparent. The PMP® is honest and does not accept inappropriate: gifts or payments or other compensation and always take the ethical high road, adheres to laws, follows the organization's policies, and is respectful of cultures, laws, and values. The project management team has a professional responsibility to its stakeholders, including customers, the performing organization, and the public. Members of PMI and also those who are PMP® certified likewise have professional responsibilities. PMI members and certified Project Management Professionals must agree to adhere to the *PMI Code of Ethics and Professional Conduct*. For more information, see: *Project Management Institute Code of Ethics and Professional Conduct*, PMI's website.

REFERENCE

A Guide to the Project Management Body of Knowledge, 4th ed., Project Management Institute, Newtown Square, PA, 2009.

Leading People Versus Managing Projects

Eldon R. Larsen, Marshall University Graduate College, South Charleston, WV

Of the many important aspects of project management, the most important key to successful project management is people. When we speak of project management, we often think of Gantt charts, time lines, budgets, resource constraints, quality challenges, performance measures, and so on. While these are important, they are neither meaningful nor effective without people. This paper presents and discusses several fundamental concepts regarding the importance of people in project management. We will discuss the importance of people in project management and their importance to project success. Emphasis will be placed on a proper understanding of communication, alignment, and teamwork.

See more of this Session: Transitioning From Engineer to Manager
See more of this Group/Topical: Management Division

Transition to Management -- A New Manager's Perspective

Stuart Everett, Eastman Chemical Company, Kingsport, TN

As someone who has recently transitioned from a technical engineering role to managment this talk will summarize my experience and feelings about this change. I'll discuss the preparation I made. How I demonstrated my ability to make the change. I'll also share my feelings after making the change, what did I expect, what surprised me, what would I prepare for.

See more of this Session: Transitioning From Engineer to Manager
See more of this Group/Topical: Management Division

Building New Relationships and Networking to Transition to Manager

Tom Mildenhall, Global Nonwovens P&TD, Kimberly-Clark Corporation, Rosewell, GA

Will be added later.

See more of this Session: Transitioning From Engineer to Manager
See more of this Group/Topical: Management Division

Communication and the Transistion to Management

Martha Nord, Nord Consultants, Nashville, TN

Communication is a vital skill for all engineers, but as an engineer plans a transition into management what is communicated to whom begins to change.

This presentation will highlight the changes in communication an engineer transitioning into management will want to make. These new communication styles and skills are looked for in people being considered for managment and are new capabilities that engineers who have become managers must show and grow.

See more of this Session: Transitioning From Engineer to Manager
See more of this Group/Topical: Management Division

Fall 2009, Nashville

Session: Transitioning from Engineer to Manager

Title: Representing the Organization as a Manager

Successful engineers are quite often tapped to become managers within organizations. The skills that made you successful as an engineer are not necessarily those needed to be a successful manager but they are a good place to start: a solid reputation for high quality work done in a timely fashion, a sense of purpose and urgency, and the willingness and recognition of the need to work with others to accomplish organizational goals.

Critical skills in management include:

- Communication Skills – within your team, across the organization and up the organization. The ability to quickly summarize the complex into a few golden nuggets (the elevator speech).
- Setting Goals and Objectives – painting a picture of where you want to take the team and obtaining the buy-in to make it happen.
- People Development – building and developing the people within your team, providing challenges and opportunities that meet their personal needs as well as helping to meet organizational goals.
- Relationships – building and adding to your network of people within the organization
- Results – ability to make decisions with incomplete data, bias for action.
- Reputation – based upon your accomplishments as well as how they were achieved

R&D Management from Beginning to End

Mukund Karanjikar
Senior Associate
Technology Holding LLC
Salt Lake City, Utah

Jack Hipple
Principal
Innovation-TRIZ
Tampa, FL

Eldon Larsen
Professor of Engineering
Marshall University
South Charleston, WV

Extended Abstract

This workshop focuses upon managing R&D projects from beginning to end. Beginning refers to mental picture of an idea whereas end refers to solving the customers need and in the process generating profit for the organization. Using a case study, the speakers will delineate challenges involved at every stage of R&D project from a mental image of an idea to a profit making product. The talk also compares old paradigm of innovation management with the new proposed paradigm. The probability of success can be significantly improved using the new paradigm.

Successful Innovation Management

We have seen many idea management flow sheets, gate management processes, and prioritization recommendations starting with an idea that already exists, but no discussion of where the idea came from or what was necessary to generate and frame it. Moreover attention is not paid to shaping and propagation of the idea. This has been especially true over the past decade when processes such as Six Sigma came into vogue whose main thrust was to eliminate deviation. New concepts and innovation are, by their very nature, deviant! In order to generate fresh and innovative ideas we must create a new thought pattern that may challenge the existing status quo in any number of ways. These can even include, at an extreme, replacing one's current products and businesses with totally new approaches and models.

The key elements of successful commercial innovation include:

1. Continuous flow of unmet needs, both from current customers and related to core competencies of the organization ("outside in" as well as "inside out"
2. Using clear mental models of problems and potential solutions
3. Perseverance for idea propagation
4. Clear definition of the business opportunity
5. An appropriate product development and management team
6. Efficient process scale up and continued input from the market
7. Successful project management to insure implementation on schedule and within budget

We will look at each one and discuss the important issues.

1. Continuous Flow of Unmet Needs and Exploration of Uses for Core Competencies

An organization needs to have a process for how it decides what to work on. This may sound simplistic, when this is left to intuition or accident; it is unlikely that the "best" projects will be chosen.

Steps to make sure that an innovation effort is capturing the best opportunities:

- Talk to groups that are not your current customers but that might be able to use the functional capability that you have as a core competence. Customer competitors are another valuable source. The customers of your customers are another excellent source. This insures that you have awareness several levels above your product and customers and alerts you to potential long term needs and trends that may not be obvious to your direct customer or may not be readily disclosed by them.

Figure 1: Level Diagram

In these discussions we want to make sure that we have a clear definition of what the "ideal result" that our customers are desiring. We can also look at parallel universes of technology and markets to explore opportunities.

2. Using Clear Mental Models of Problems and Solutions

It is one thing to "invent" a new product, but another to make sure that we understand how a potential customer is going to use the product as well as how the customer's customer is going to use their product. The days are gone when we can just make something, put it in a drum or tank car, and wait for the invoice to be paid. We must be intimately familiar with the product's use, its limitations (including regulatory issues if appropriate). We must understand the details of the physical use of the product and how it affects a customer's products and business.

It is also important to make sure that we do not have blinders on when thinking about use of our product. It is rare that there is only one way to accomplish a function. We must be constantly vigilant about competitive activities, not just from direct competitors, but from parallel approaches to meeting a customer's needs.

3. Perseverance for Idea Propagation

New business and product development is hard work and it is easy to get discouraged. An idea that starts our with a 10% probability of success can raise its odds to over 75% by the proper, early stage identification of what challenges must be overcome. If these are not clearly defined, then a new product or business development will flounder and struggle. The perseverance battle rests more on the shoulders of individuals than the organization or top management. Idea has to propagate and the best person to do so is the idea owner. Ideas may be facilitated top-down (by management) however propagation is always bottom up (by employees). A proprietary tool (YouSNAP®) is being proposed to better manage idea propagation. The five elements of the tool are:
a. You – Know yourself
b. Surroundings – Know your surroundings
c. Network – Develop your network
d. Approach – Decide an approach
e. Persistence – Be persistent and iterate

Employees need to understand a simple truth. Ideas are dime a dozen and never bought on face value. Frustration is part of innovation. One should know that failures will exist. Hence know yourself. Are you up to it? Then you should know your surroundings. Often times, a tendency exists in corporations to blame the corporation itself. What is corporation? It is made up of people like yourself trying to serve the customers. If there are non-cooperative people around you, the journey would be commensurately difficult. However the important thing is that you should know this before. This leads to having a robust network of supporters. Once you have done the necessary preparation, decide upon an approach. Finally be persistent. The presentation will discuss the tool in more interactive format.

4. The Business Opportunity

A new product is not a business unless we have a way of sustaining the product's sales, expanding its utility, and developing and enforcing an intellectual property portfolio. As we commercialize a new product, we must be constantly aware of, and be willing to address with resources and commitment, issues such as change in product form, packaging and shipping issues, product purity, and shelf life/stability. These issues are most adequately addressed by constant and continuous communication with the customer.

5. Product Development Teams

We frequently choose product team members based on "who's available". While this may be expedient, it may not serve the best interests of the new business development. The obvious component of team selection is function—marketing R&D, manufacturing, legal, sales, etc. but we frequently ignore the human side of team functioning. If team members spend too much time dealing with personal issues relating to other team members, that energy is not applied to successful new product launch. Tools that can assist in this include measurements such as the Myers Briggs family of instruments, the Hermann Brain Dominance instrument (HBDI), and the Kirton KAI assessment. Knowledge of these personality profiles will not change natural behavior, but can make team members aware of others' preferred styles and open discussion can eliminate hidden barriers to effective team functioning.

6. Efficient process scale up and continued input from the market

Often times, companies strive for launching the best product. It is important to realize that continued input from the market is critical in order to have that 'best product' in the market. It is necessary to efficiently scale up and launch a product sooner than later. It should be perfected using market feedback. Getting market feedback is the only way to close the R&D loop and serve your customers to the fullest.

7. Successful Project Implementation and Management

New product development does not end when the first semi-commercial sample has been OK'd by the customer. A new type of team is now required to oversee the details process design, plant construction (including all the sub issues of permitting, EPA regulations, etc.). Quick and efficient plant start up with appropriate contingency planning is also a part of this effort. Contingency planning, not only related to unforeseen plant construction and start up difficulties, but also to taking care of the customer's needs in these situations must be part of the plan.

Three keys to successful project management lead to successful project completion. These keys involve

- The importance of people
- The importance of defining and setting good objectives
- The importance of good planning

We will apply these three keys to successful project management. This is all about application, not just theory. Today, you can apply the lessons you will learn.

A case study will be discussed in the presentation.

Industry Perspective On Sponsored Research

Gretchen Baier, Core R&D, The Dow Chemical Company, Midland, MI

Many factors contribute to a successful sponsored research collaboration among industry, government and universities. A few critical issues get in the way. The conflicting views, interests, and policies across the various participants, including the university technology transfer offices, the sponsoring company, the governmental organizations, and the individual professor, regarding rights to inventions are sources of constant friction that significantly limits industry's willingness to sponsor university research. The Dow Chemical Company has more than twenty years of experience in centrally coordinating a wide variety of successful research collaborations, both in the U.S. and abroad. This presentation will give an industry perspective on these issues in the hope that a better understanding of the needs and drivers in the mature manufacturing industries will lead to more flexibility among the various stakeholders.

See more of this Session: Managing Technology Collaborations -- Industry, Academia and Government
See more of this Group/Topical: Management Division

Traversing the University-Industry Interface--Overcoming the Challenges

Robert R. Gruetzmacher, DuPont, Wilmington, DE

The title of this session is an expression of the interest on the part of companies and universities to work together to create business opportunity. Universities are becoming key sources of discovery and with recent changes in government funding (amount and process) academia will be playing an ever increasing role in how we innovate. Global competition, rising R&D costs and thus the need to get more products to the market sooner are some factors forcing companies to reach beyond themselves, e.g. to research universities, for new ideas and capability. Bridging the traditional divide between the academic and company missions is a challenge but one that is being overcome. Communities, some with government backing, have formed to brainstorm and benchmark through the challenges. The University-Industry-Demonstration Partnership (UIDP, www.UIDP.org) is a good example. Successful university/company engagement is recognized to be in the National interest. This talk will describe what some companies and universities that are serious about working together are doing to traverse this interface effectively.

See more of this Session: Managing Technology Collaborations -- Industry, Academia and Government
See more of this Group/Topical: Management Division

Developing AIChE National Student Design Competition Problems by Collaborations Among Industry and Government Sponsors and Academia Participants

Richard L. Long, New Mexico State University, Las Cruces, NM

The AIChE National Student Design Competition has been held annually since 1932. This problem is distributed to every department of chemical engineering in the U.S. and the competition is open to all who comply with its ruless. Typically, 20-30 schools out of 157 dpartments of ChE in the U.S. participate in the competition and a roughly equal number use the problem in their senior design sequence without participating in the competition. The search for sponsors is an ongoing task. The appeal to students of industrial and governmental sponsors is that the problems are often considered more "real" than those that might be provided through academics alone. The challenge is to attract industrial and government sponsors to assist with the preparation of the problem and the evaluation of the national competition.

See more of this Session: Managing Technology Collaborations -- Industry, Academia and Government
See more of this Group/Topical: Management Division

Conquering Innovation Fatigue in University-Industry Collaboration: Recommendations for Success

Jeff Lindsay[1], Cheryl Perkins and Mukund Karanjikar

There is widespread recognition today that IP (intellectual property) from university research is a rich source of potential value to be mined. A few schools have seen revenues of hundreds of millions of dollars for some of their patents, inspiring hopes of finding the next Gatorade®. However, some university efforts to extract value from their IP may have unintended consequences that limit the potential of university-industry collaboration. It's important to understand the potential barriers and steps that can lead to more successful collaboration.

Some challenges stem from the demands or misapplications of the Bayh-Dole Act, or the Patent and Trademark Law Amendments Act[1] that was signed into law December 12, 1980. This law was intended to strengthen university-industry cooperation by, among other things, permitting universities to own and license patents that stem from federal grants. While the Act had many positive benefits, there may be some

---

[1] Contact and presenter: Jeff Lindsay, Director of Solution Development, Innovationedge LLC, 1526 South Commercial Street #200, Neenah, WI 54956. jlindsay@innovationedge.com or inventmore@gmail.com.

unintended consequences which are driving some companies away from U.S. universities and toward foreign universities or away from university collaboration altogether.

Companies sometimes complain that universities use the Bayh-Dole Act to justify their requirement of owning all IP from privately-funded research, although the Act provides for an exception for such research (Section 401.1).[2] (Also see the testimony of Dr. Susan Butts on this issue.[3,4]) Many companies balk at the IP-related demands of many universities, feeling that it is unreasonable to be denied IP rights for work they sponsor. Much more flexible IP provisions can be negotiated with many foreign schools, contributing to a flux of US research money going oversees.

Some of the increased tension between universities and industry (at least in some sectors) may not be not inherent to the Bayh-Dole Act, nor is it necessarily due to unreasonable expectations of universities or prospective industry partners. Some of the challenges may be exacerbated by tax regulations, coupled with misaligned incentives in typical universities. Solutions may require enlightened legislators and officials adjusting tax regulations and university officials creating aligned objectives and incentives. Industries can also adjust their expectations and look to more open models of collaboration with universities for improved results where IP rights aren't the dominant outcome.

Particular challenges arise from IRS tax regulations, specifically the policies known as Rev. Proc. 1997-14[5] and Rev. Proc. 2007-47.[6] These rules prove "safe-harbor" guidelines suggesting that the tax-exempt status of a bond to fund university buildings could be at risk if privately funded research for "private benefit" exceeds a low threshold of 5-10%. These guidelines are generally interpreted by universities in their favor, of course. The implications have resulted in substantial delays in negotiating contracts with industry, and these problems are often incorrectly blamed on the Bayh-Dole Act.

These problems are further compounded by poor alignment of incentives in universities. Tech transfer offices may have financial incentives to seek the highest possible returns on licensed technology, whereas professors themselves may have little incentive to pursue IP. University leaders may have little incentive to work with industry unless it provides massive funding (say, over $50 million). All these conflicting interests can stall effective relationships with industry. (These problems and prospective solutions are discussed in more detail in *Conquering Innovation Fatigue*, the 2009 publication from John Wiley & Sons by the present authors.)

In corporations, there are also challenges as the concept of "open innovation" can be viewed with suspicion and anxiety. A champion is often needed who can manage the hurdles involved with university collaboration and help the organization understand the value that can be delivered. Such champions need to be well connected with business groups and benefit from also having an academic background or being close to academia so they can related effectively to professors and others in the university and manage the varied expectations and needs.

Fortunately, there are some energizing factors that can overcome some of the "innovation fatigue factors" we discuss in this presentation. Some universities offer generous portions of the royalties from patents back to the professors (e.g., 25-60%). Some also provide assistance to start-ups and provide other encouragements to help new businesses to arise from work originating at the university. Such factors can be powerful incentives for innovation that help foster ongoing industry-university collaboration.

Notes:

1. P.L. 96-517, codified at 35 USC § 200 et seq.

2. 37CFR401.1, Code of Federal Regulations, Title 37, Volume 1, U.S. Government Printing Office, Washington, DC, revised as of July 1, 2002, pp. 609-610, http://edocket.access.gpo.gov/cfr_2002/julqtr/37cfr401.1.htm.

3. Susan Butts, "Federal Investment in Research: Implications for U.S. Competitiveness–An Industry Perspective," PCAST Meeting, Aug,. 24, 2007, slide 7, slides available online at www.erc-assoc.org/annmtg/2007_meeting_files/ILO_Susan%20Butts.pdf (or http://tinyurl.com/cvejk2) (accessed April 10, 2008).

4. Susan Butts, "Testimony Before the Subcommittee on Technology and Innovation, Committee on Science and Technology, U.S. House of Representatives: Bayh-Dole – The Next 25 Years," July 17, 2007, Washington, DC, http://democrats.science.house.gov/Media/File/Commdocs/hearings/2007/tech/17jul/butts_testimony.pdf (or http://tinyurl.com/bgxw7g) (accessed March 7, 2009).

5. Revenue Procedure 1997-14, Internal Revenue Service, 1997, available online at www.unclefed.com/Tax-Bulls/1997/Rp97-14.pdf, (accessed April 10, 2008).

6. Revenue Procedure 2007-47, Internal Revenue Service, 2007, www.irs.gov/pub/irs-drop/rp-07-47.pdf (accessed April 10, 2008).

EXTENDED ABSTRACT

Christopher D. McKinney, Vanderbilt University

Improving Collaborations: The Keyhole Theory of Management and Other Reflections

as part of Management Division
Session: Managing Technology Collaborations -- Industry, Academia and Government
scheduled for Wednesday, November 11, 2009: 8:30 AM-11:00 AM

Industry, academia, and government should work together ever more closely in conceiving, developing, and commercializing promising new innovations. They each bring special strengths that can multiply effectiveness. In essence, these three players have important opportunities for making "the whole greater than the sum of the parts." So, what are critical elements to creating this innovation Gestalt effect? We'll talk about key drivers for each player, as well as important constraints --- and find some common ground upon which the three can build together. We'll also review important "working together" points that, while often taught more in the negotiation context, are readily applicable in the day-to-day context of doing business. Finally, we'll spend time on the "keyhole theory of management," a concept focused on the effective understanding of perception based on limited data and the opportunities that are created by this "keyhole" for greater success in industry-academia-government collaborations.

AUTHOR INDEX

American Institute of Chemical Engineers
3 Park Avenue
New York, New York 10016-5991